Electrical and Electro-Optical Biosensors

Electrical and Electro-Optical Biosensors

Editors

Mon-Juan Lee
Seunghyun Kim

MDPI • Basel • Beijing • Wuhan • Barcelona • Belgrade • Manchester • Tokyo • Cluj • Tianjin

Editors

Mon-Juan Lee
Chang Jung Christian University
Taiwan

Seunghyun Kim
Baylor University
USA

Editorial Office
MDPI
St. Alban-Anlage 66
4052 Basel, Switzerland

This is a reprint of articles from the Special Issue published online in the open access journal *Biosensors* (ISSN 2079-6374) (available at: https://www.mdpi.com/journal/biosensors/special_issues/elect_optical_biosens).

For citation purposes, cite each article independently as indicated on the article page online and as indicated below:

LastName, A.A.; LastName, B.B.; LastName, C.C. Article Title. *Journal Name* **Year**, *Volume Number*, Page Range.

ISBN 978-3-0365-3989-8 (Hbk)
ISBN 978-3-0365-3990-4 (PDF)

© 2022 by the authors. Articles in this book are Open Access and distributed under the Creative Commons Attribution (CC BY) license, which allows users to download, copy and build upon published articles, as long as the author and publisher are properly credited, which ensures maximum dissemination and a wider impact of our publications.

The book as a whole is distributed by MDPI under the terms and conditions of the Creative Commons license CC BY-NC-ND.

Contents

About the Editors vii

Mon-Juan Lee
A Label-Free and Affordable Solution to Point-of-Care Testing Devices
Reprinted from: *Biosensors* **2022**, *12*, 192, doi:10.3390/bios12040192 1

Ina Turcan, Iuliana Caras, Thomas Gabriel Schreiner, Catalin Tucureanu, Aurora Salageanu, Valentin Vasile, Marioara Avram, Bianca Tincu and Marius Andrei Olariu
Dielectrophoretic and Electrical Impedance Differentiation of Cancerous Cells Based on Biophysical Phenotype
Reprinted from: *Biosensors* **2021**, *11*, 401, doi:10.3390/bios11100401 5

Fahmida Nasrin, Kenta Tsuruga, Doddy Irawan Setyo Utomo, Ankan Dutta Chowdhury and Enoch Y. Park
Design and Analysis of a Single System of Impedimetric Biosensors for the Detection of Mosquito-Borne Viruses
Reprinted from: *Biosensors* **2021**, *11*, 376, doi:10.3390/bios11100376 17

Po-Chang Wu, Chao-Ping Pai, Mon-Juan Lee and Wei Lee
A Single-Substrate Biosensor with Spin-Coated Liquid Crystal Film for Simple, Sensitive and Label-Free Protein Detection
Reprinted from: *Biosensors* **2021**, *11*, 374, doi:10.3390/bios11100374 31

Yelim Kim, Ahmed Salim and Sungjoon Lim
Millimeter-Wave-Based Spoof Localized Surface Plasmonic Resonator for Sensing Glucose Concentration
Reprinted from: *Biosensors* **2021**, *11*, 358, doi:10.3390/bios11100358 45

Hassanein Shaban, Mon-Juan Lee and Wei Lee
Label-Free Detection and Spectrometrically Quantitative Analysis of the Cancer Biomarker CA125 Based on Lyotropic Chromonic Liquid Crystal
Reprinted from: *Biosensors* **2021**, *11*, 271, doi:10.3390/bios11080271 59

Hassanein Shaban, Shih-Chun Yen, Mon-Juan Lee and Wei Lee
Signal Amplification in an Optical and Dielectric Biosensor Employing Liquid Crystal-Photopolymer Composite as the Sensing Medium
Reprinted from: *Biosensors* **2021**, *11*, 81, doi:10.3390/bios11030081 73

Kristina A. Malsagova, Tatyana O. Pleshakova, Rafael A. Galiullin, Andrey F. Kozlov, Ivan D. Shumov, Vladimir P. Popov, Fedor V. Tikhonenko, Alexander V. Glukhov, Vadim S. Ziborov, Oleg F. Petrov, Vladimir E. Fortov, Alexander I. Archakov and Yuri D. Ivanov
Highly Sensitive Detection of CA 125 Protein with the Use of an n-Type Nanowire Biosensor
Reprinted from: *Biosensors* **2020**, *10*, 210, doi:10.3390/bios10120210 89

Emilio Sardini, Mauro Serpelloni and Sarah Tonello
Printed Electrochemical Biosensors: Opportunities and Metrological Challenges
Reprinted from: *Biosensors* **2020**, *10*, 166, doi:10.3390/bios10110166 97

Hamed Shamkhalichenar, Collin J. Bueche and Jin-Woo Choi
Printed Circuit Board (PCB) Technology for Electrochemical Sensors and Sensing Platforms
Reprinted from: *Biosensors* **2020**, *10*, 159, doi:10.3390/bios10110159 127

About the Editors

Mon-Juan Lee received her bachelor's and master's degrees in chemical engineering in 1998 and 2000, respectively, and a PhD degree in life science in 2006 from National Tsing Hua University, Hsinchu, Taiwan. She is currently the director of the Department of Medical Science Industries and a professor of the Department of Bioscience Technology at Chang Jung Christian University, Tainan, Taiwan. Her research interests include bone-stimulating drug development and drug screening, nanobiomaterials as nonviral vectors for gene delivery, as well as LC-based biosensing technologies. She received the Special Outstanding Talent Award from the Ministry of Science and Technology, Taiwan, in 2014, 2016, and 2018.

Seunghyun Kim, Ph.D., is an associate professor of electrical and computer engineering at Baylor University. He oversees the Bio and Micro Devices Lab (BMDL) and a class 1,000 cleanrooms at Baylor Research and Innovation Collaborative, a state-of-the-art research facility at Baylor University. His extensive research work in biosensors, integrated optics, photonic crystals, and micro/nanofabrication has led to 4 patents, 37 refereed journal papers, and 57 conference abstracts and proceedings. Dr. Kim received a National Science Foundation (NSF) CAREER award in 2014 and Baylor's Outstanding Professor Award for a distinctive scholarship in 2018. He was selected as a rising star at Baylor University in 2017. He has been a reviewer for various journals in optics and biomedical fields and has also served as a panel/ad hoc reviewer for the NSF and the National Institutes of Health. Dr. Kim is a senior member of Optica (formerly OSA) and a member of the IEEE and BMES.

Editorial

A Label-Free and Affordable Solution to Point-of-Care Testing Devices

Mon-Juan Lee [1,2]

[1] Department of Bioscience Technology, Chang Jung Christian University, Tainan 71101, Taiwan; mjlee@mail.cjcu.edu.tw
[2] Department of Medical Science Industries, Chang Jung Christian University, Tainan 71101, Taiwan

Citation: Lee, M.-J. A Label-Free and Affordable Solution to Point-of-Care Testing Devices. *Biosensors* **2022**, *12*, 192. https://doi.org/10.3390/bios12040192

Received: 18 March 2022
Accepted: 22 March 2022
Published: 24 March 2022

Publisher's Note: MDPI stays neutral with regard to jurisdictional claims in published maps and institutional affiliations.

Copyright: © 2022 by the author. Licensee MDPI, Basel, Switzerland. This article is an open access article distributed under the terms and conditions of the Creative Commons Attribution (CC BY) license (https://creativecommons.org/licenses/by/4.0/).

Clinical diagnosis and disease monitoring often require the detection of small-molecule analytes and disease-related proteins in body fluids. Most conventional biochemical assays for protein detection are label-based immunoassays such as the enzyme-linked immunosorbent assay (ELISA), the gold standard of immunoassays. Fluorescence-labeled antibodies are commonly used in these bioanalytical methods to signal the formation of immunocomplexes as a result of the specific binding between an antigen and an antibody, either of which can be the target of detection. Both fluorescence labeling and fluorometric instruments increase the cost of the analysis, and limit the operation of such assays to trained medical professionals. In an era of global pandemics, the need for personal health monitoring through point-of-care testing (POCT) devices, especially during home quarantine, has become even more imperative. To develop POCT devices with lower cost that can be used more prevalently among untrained personnel, label-free detection approaches, such as electrical and electro-optical biosensing, are being investigated in the hope of enhancing detection sensitivity and increasing the linear range of detection. This Special Issue, "Electrical and Electro-Optical Biosensors", includes six research articles covering biosensors based on electrochemical impedance, localized surface plasmon resonance, and dielectric properties of liquid crystals (LCs), as well as two review articles on printed electrochemical biosensors. The reported biosensors were designed to detect glucose, bovine serum albumin (BSA), and the cancer biomarker CA125; and to discern between different mosquito-borne viruses and cancer cells.

Surface plasmon resonance (SPR), the electron oscillation stimulated by an incident light on metal surface, is one of the most common label-free biosensing technologies. Localized SPR (LSPR) refers to SPR confined to nanoparticles with diameters comparable to the wavelength of the incident light, which is suggested to enhance detection sensitivity through signal amplification. A millimeter-wave-based spoof localized surface plasmonic resonator was developed for glucose detection in a microfluidic system [1]. The millimeter-wave-based glucose sensor exhibited higher sensitivity than microwave-based sensors with a limit of detection (LOD) of 1 mg/dL from a sample volume of 3.4 µL, and is reusable with satisfactory reproducibility. The performance of the LSPR glucose sensor is comparable to commercial glucose sensors such as the Accu-Chek blood glucose meters manufactured by Roche, which rely on electrochemical signals produced by the reaction of glucose with glucose oxidase and have a detection range of 10–600 mg/dL glucose for a sample volume ranging from 0.3 to 2 µL.

Electrical biosensors detect biological binding events occurring on the electrode and transduce electrical signals in the form of conductance, resistance, or capacitance, which are dependent on the amount of analyte. In an electrical biosensor consisting of a silicon-on-insulator nanowire immobilized with antibodies against CA125, the dependence of electric current on CA125 concentration was established [2]. The high surface-to-volume ratio of the nanowire enabled sensitive detection of CA125 to concentrations as low as 10^{-16} M. On the other hand, electrochemical impedance spectroscopy (EIS) has become a powerful tool for electrical biosensing. Biomolecular recognition resulted from antigen–antibody or ligand–receptor binding alters the electron transfer resistance at the surface of the electrode, which can then be quantitatively analyzed and correlated to the amount of analytes. An impedimetric biosensor based on a gold–polyaniline and sulfur-/nitrogen-doped graphene quantum dot nanocomposite conjugated with antibodies was designed for the detection of mosquito-borne viruses [3]. Dengue virus, zika virus, and chikungunya virus were discerned by the impedimetric biosensor with minimal cross-reactivity and LOD in the range of femtogram per milliliter. Moreover, the dielectric characteristics of several cancer cell types, represented by crossover frequencies, were determined by capturing cells on interdigitated microelectrodes through dielectrophoresis, followed by EIS analysis [4]. These results demonstrate that the selectivity or specificity of detection in electrical biosensors can be realized by including target-specific antibodies in the sensor design, or by examining the unique electrical signal produced by the analyte.

Printed circuit board (PCB) is one of the key technologies to miniaturize and lower the cost of point-of-care testing devices. It also facilitates easy integration of the sensing platform with more sophisticated electronic and microfluidic systems [5,6]. Consisting of multilayers of conductive and insulating materials, PCB was originally a component of the integrated circuit for the electronics industry [6]. In electrochemical biosensors, which transduce biochemical signals through amperometric, impedimetric, or potentiometric principles, the two- or three-electrode circuit system was printed on a small surface area by screen printing, inkjet printing, or aerosol jet printing procedures, which varies in resolution and ink dispensing methods [5]. Various printing strategies were established to enhance detection sensitivity and LOD, as well as to increase biocompatibility for the purpose of direct detection in a biological environment. With a growing demand for medical wearable devices, the increasing versatility of PCB technology was seen in the development of flexible and stretchable PCBs. Currently, printed electrochemical biosensors utilizing amperometry, cyclic voltammetry, and EIS were developed for the detection of small-molecule metabolites such as glucose and lactate, disease-related marker proteins such as interferon-gamma, DNA associated with single nucleotide polymorphism, and whole cells such as eukaryotic cells and pathogens [5,6]. With novel printing technology and fabrication procedures to improve metrological performance, including signal-to-noise ratio, detection sensitivity, repeatability, and reproducibility, PCB-based biosensors are expected to become the mainstream biosensing technologies for affordable point-of-care diagnostics.

LCs have become an indispensable material in our daily lives, seen predominantly in LC display devices such as smartphones, digital clocks, and flat-screen televisions. LCs are fluidic but exhibit molecular order similar to solid crystals, and can be induced to reorient their molecular alignment under the influence of temperature, electromagnetic radiation, electric or magnetic fields. Because the optical, electrical, and electro-optical properties of LCs are altered in a concentration-dependent manner by biological analytes, biosensing application of LCs has been extensively explored in recent decades. Conventional LC-based biodetection at the LC–glass interface was performed in a LC cell with a thin film of LCs sandwiched between a pair of glass substrates. By doping the nematic LC E7 with a prepolymer, NOA65, followed by photopolymerization to produce a LC–photopolymer composite, the optical and dielectric signal of the LC-based biosensor can be enhanced [7]. To simplify the preparation procedure of the LC–glass detection platform, a single glass substrate spin-coated with a LC film, instead of a LC cell, was utilized in the detection of BSA and CA125 [8]. Signal amplification was achieved in such single-substrate detection due to the reduced film thickness of the spin-coated LC film. Most LC-based biosensors reported to date consist of thermotropic LCs, which dominate the LC display industry. Nevertheless, lyotropic LCs, which are hydrophilic and thus more biocompatible, may hold greater potential in the biomedical application of LCs. Biosensing techniques based on the nematic phase of disodium cromoglycate (DSCG), a type of lyotropic chromonic LCs, were demonstrated in the detection of BSA and CA125 with a LOD comparable to those of nematic thermotropic LC-based biosensors [9].

Electrical and electro-optical biosensing technologies are critical to the development of innovative POCT devices, which can be used by both professional and untrained personnel to provide necessary health information within a short time for medical decision to be made, and are especially important in an era of global pandemics. This Special Issue includes some of the pioneering work on biosensors utilizing electrochemical impedance, localized surface plasmon resonance, and bioelectricity of sensing materials as the signal response that is pertinent to the amount of analyte. The results presented demonstrate the potential of these label-free biosensing approaches in the detection of disease-related small-molecule metabolites, proteins, and whole-cell entities.

Funding: This research received no external funding.

Acknowledgments: The author is grateful for the opportunity to serve as one of the guest editors of the Special Issue, "Electrical and Electro-Optical Biosensors", as well as the contribution of all the authors to this Special Issue. The dedicated work of Special Issue Editor of Biosensors, and the editorial and publishing staff of Biosensors is greatly appreciated.

Conflicts of Interest: The author declares no conflict of interest.

References

1. Kim, Y.; Salim, A.; Lim, S. Millimeter-Wave-Based Spoof Localized Surface Plasmonic Resonator for Sensing Glucose Concentration. *Biosensors* **2021**, *11*, 358. [CrossRef] [PubMed]
2. Malsagova, K.A.; Pleshakova, T.O.; Galiullin, R.A.; Kozlov, A.F.; Shumov, I.D.; Popov, V.P.; Tikhonenko, F.V.; Glukhov, A.V.; Ziborov, V.S.; Petrov, O.F.; et al. Highly Sensitive Detection of CA 125 Protein with the Use of an n-Type Nanowire Biosensor. *Biosensors* **2020**, *10*, 210.
3. Nasrin, F.; Tsuruga, K.; Utomo, D.I.S.; Chowdhury, A.D.; Park, E.Y. Design and Analysis of a Single System of Impedimetric Biosensors for the Detection of Mosquito-Borne Viruses. *Biosensors* **2021**, *11*, 376. [PubMed]
4. Turcan, I.; Caras, I.; Schreiner, T.G.; Tucureanu, C.; Salageanu, A.; Vasile, V.; Avram, M.; Tincu, B.; Olariu, M.A. Dielectrophoretic and Electrical Impedance Differentiation of Cancerous Cells Based on Biophysical Phenotype. *Biosensors* **2021**, *11*, 401. [CrossRef] [PubMed]
5. Sardini, E.; Serpelloni, M.; Tonello, S. Printed Electrochemical Biosensors: Opportunities and Metrological Challenges. *Biosensors* **2020**, *10*, 166.
6. Shamkhalichenar, H.; Bueche, C.J.; Choi, J.-W. Printed Circuit Board (PCB) Technology for Electrochemical Sensors and Sensing Platforms. *Biosensors* **2020**, *10*, 159. [CrossRef]
7. Shaban, H.; Yen, S.-C.; Lee, M.-J.; Lee, W. Signal Amplification in an Optical and Dielectric Biosensor Employing Liquid Crystal-Photopolymer Composite as the Sensing Medium. *Biosensors* **2021**, *11*, 81. [PubMed]
8. Wu, P.-C.; Pai, C.-P.; Lee, M.-J.; Lee, W. A Single-Substrate Biosensor with Spin-Coated Liquid Crystal Film for Simple, Sensitive and Label-Free Protein Detection. *Biosensors* **2021**, *11*, 374. [PubMed]
9. Shaban, H.; Lee, M.-J.; Lee, W. Label-Free Detection and Spectrometrically Quantitative Analysis of the Cancer Biomarker CA125 Based on Lyotropic Chromonic Liquid Crystal. *Biosensors* **2021**, *11*, 271. [CrossRef] [PubMed]

Article

Dielectrophoretic and Electrical Impedance Differentiation of Cancerous Cells Based on Biophysical Phenotype

Ina Turcan [1,†], Iuliana Caras [2,†], Thomas Gabriel Schreiner [1,3], Catalin Tucureanu [2], Aurora Salageanu [2], Valentin Vasile [2], Marioara Avram [4,5], Bianca Tincu [4,6] and Marius Andrei Olariu [1,*]

1. Department of Electrical Measurements and Materials, Faculty of Electrical Engineering and Information Technology, Gheorghe Asachi Technical University of Iasi, 21-23 Profesor Dimitrie Mangeron Blvd., 700050 Iasi, Romania; ina.turcan@student.tuiasi.ro (I.T.); thomas.schreiner@umfiasi.ro (T.G.S.)
2. "Cantacuzino" National Medical-Military Institute for Research and Development, 103 Splaiul Independentei, 050096 Bucharest, Romania; caras.iuliana@cantacuzino.ro (I.C.); tucureanu.catalin@cantacuzino.ro (C.T.); salageanu.aurora@cantacuzino.ro (A.S.); vasile.valentin@cantacuzino.ro (V.V.)
3. Faculty of Medicine, "Grigore T. Popa" University of Medicine and Pharmacy, 16 Universitatii Street, 700115 Iasi, Romania
4. National Institute for Research and Development in Microtechnologies—IMT Bucharest, 126A Erou Iancu Nicolae Street, 077190 Bucharest, Romania; marioara.avram@imt.ro (M.A.); bianca.tincu@imt.ro (B.T.)
5. DDS Diagnostic SRL, 7 Vulcan Judetu Street, 030423 Bucharest, Romania
6. Faculty of Applied Chemistry and Material Science, University "Politehnica" of Bucharest, 313 Splaiul Indepentei, 060042 Bucharest, Romania
* Correspondence: molariu@tuiasi.ro; Tel.: +40-744-474-232
† These authors equally contributed to this work.

Abstract: Here, we reported a study on the detection and electrical characterization of both cancer cell line and primary tumor cells. Dielectrophoresis (DEP) and electrical impedance spectroscopy (EIS) were jointly employed to enable the rapid and label-free differentiation of various cancer cells from normal ones. The primary tumor cells that were collected from two colorectal cancer patients, cancer cell lines (SW-403, Jurkat, and THP-1), and healthy peripheral blood mononuclear cells (PBMCs) were trapped first at the level of interdigitated microelectrodes with the help of dielectrophoresis. Correlation of the cells dielectric characteristics that was obtained via electrical impedance spectroscopy (EIS) allowed evident differentiation of the various types of cell. The differentiations were assigned to a "dielectric phenotype" based on their crossover frequencies. Finally, Randles equivalent circuit model was employed for highlighting the differences with regard to a series group of charge transport resistance and constant phase element for cancerous and normal cells.

Keywords: cancer cells; dielectrophoresis; crossover frequency; electrical impedance spectroscopy

1. Introduction

Label-free manipulation, sorting, and isolation of biological cells and, in particular, cancerous cells, are a matter of concern at a worldwide level. Dielectrophoresis (DEP) has emerged as a potential technique for this purpose since early 90's. To date, DEP has been extensively employed in the electromanipulation of cancer cells and many studies can be provided as examples of best practices. The increased interest in DEP-based technique utilization is justified mainly by the fact that the technique does not require prior knowledge of specific cells, as it does in biomarker related isolation techniques. Capturing cells on DEP systems has the enormous advantage of reversibility, maintaining the cells viability for further characterization and culturing. Initial experimental DEP-based studies focused on the manipulation, sorting, and isolation of cancer cells. DEP experiments that focused on various types of cancer cells have been already reported: breast, MCF-10A, MCF-7, MDA-MB-231 and MDA-MB-435 [1–3]; oral, HOK, H357, H157 [4,5]; leukemia, K526 [6]; kidney,

HEK 293, 786-O [7,8]; ovary, SKOV-3 [3,9]; prostate PC3, LnCap [3]; lung A549, H1299, 95C, 95D [10,11]; cervical, HeLa [12]; and colorectal, HCT-116 [2]. The aforementioned studies highlight the possibility of discriminating cancerous cells on the basis of their crossover frequencies. Recently, Turcan and Olariu [13] presented in a centralized manner, the evolution of dielectric parameters versus crossover frequencies.

On the other hand, biophysical characterization of cancer cells on the basis of impedance measurements has been studied with the aim of identifying the "electrical signature" of various types of cancers, which may allow for the label-free successful evaluation of therapeutic efficiency. Both the impedance flow cytometry (IFC) and the electrical impedance spectroscopy (EIS) were used as techniques to gather impedance data for bulk cell suspensions, clustered cells, or single cells. EIS is characterizing the evolution of dielectric parameters against the frequency as a result of the interaction between an electrical stimulus (i.e., external electric field) and the biological matter. The dielectric behaviour of polarized cells is analysed with respect to the evolution of three (α-, β-, and γ-) dispersions of different magnitudes which may occur at different frequencies.

Human cancers consist of cells that display different phenotypic features, including cellular morphology, gene expression, metabolism, motility, proliferation, and metastatic potential [14]. This heterogeneity is a result of the interplay between cell-intrinsic (i.e., the variability in the genetics, epigenetics, and the biology of a tumor's cell-of-origin) and cell-extrinsic factors (i.e., those arising from factors in the microenvironment) which shape the cellular phenotype [15,16]. Consequently, the phenotypic heterogeneity within tumors constitutes a major impediment in their diagnostics and therapy. From this point of view, EIS has the ability to monitor the dynamics of intrinsic and extrinsic changes that occur in cancer cells [17].

Breast mammalian cancer cells are among the most studied cancer cells in dielectric studies [18–21]. Qiao et al. [18] employed electrical impedance spectroscopy for monitoring a cells' state in solutions. The measurements were performed between 300 kHz and 1 MHz at the level of MDA-MB-435S, MDA-MB-231, MDA-MB-7, and MCF-10A cell lines. The characteristic relaxations increased from normal to late cancer stages which allowed clear differentiation of each cell's electrical signature. Moreover, Huerta-Nuñez reported the [19] successful identification of breast cancer with the help of impedance spectroscopy by performing studies on solutions of non-metastatic (MCF-7, MDA-MB-231) and metastatic (SK-BR-3) breast cancer cells that were coupled with magnetic nanoparticles of very low concentrations.

The comparative impedance measurements on lung and liver cancer cells were reported by Al Ahmad [22] who highlighted the reduced ability of cancerous cells for storing energy in comparison to normal cells. In [23], Zhang reported not only the capability of impedimetric measurements for distinguishing between skin cancer cells (A431) and normal cells (HaCaT), but also confirmed the capacity of the technique of providing real-time kinetic information on cell proliferation behaviour.

Therefore, electromanipulation and electrical characterization of cancerous cells has demonstrated good differentiation among various types of cells from an electrical viewpoint. A much more powerful tool may be developed by combining EIS with DEP. Nguyen et al. concentrated A549 cells while applying p-DEP (positive DEP) at the level of circular microelectrodes at the frequency of 1 MHz and the potential of 10 Vpp (peak-to-peak voltage) [10]. The impedance measurements that were performed demonstrated a linear relation between the impedance variation and the cells' number, and therefore, the high potential of the technique for being employed when there is a low quantity of cells. Thus, combined exploitation of EIS and DEP provides supplementary information on cancer cell dielectric properties and correlations to their biophysical phenotype can be made.

In this paper, we are reporting on the utilization of DEP for trapping cancer cell lines as well as primary tumor cells. All the cells were firstly suspended in a low conductivity suspension medium and concentrated with the help of dielectrophoresis at the level of interdigitated (castellated) microelectrodes. The differentiation among the different types of cancer cells (including primary tumor cells that were collected from two colorectal cancer patients and cancer cell lines (SW-403, Jurkat and THP-1)), and healthy peripheral blood mononuclear cells (PBMCs) was done based on EIS experiments following DEP cells' trapping and identification of crossover frequencies for each type of cell.

2. Materials and Methods

2.1. Fabrication of Interdigitated Microelectrodes

The interdigitated microelectrodes were manufactured within a clean room facility class 1000 (ISO 5). Metal-based microelectrodes were fabricated by a lift-off technique on an oxidized 4-inch silicon (Si) wafer, using photoresist as a sacrificial layer. LOR 10B photoresist was spin coated on top of the Si wafer at 3000 rpm for 30 s and pre-baked on a hot plate at 150 °C for 5 min, followed by a spin coating of HPR 504 photoresist at 2000 rpm for 30 s and pre-baking on a hot plate at 95 °C for 45 s. The two photoresists were imprinted by UV lithography; exposure was performed in a MA6 mask aligner (Suss MicroTec) for 2.5 s to transfer the pattern from the photolithographic mask to the photoresist. Following UV exposure, the photoresist was developed in a specific solution (HPRD 402) for 30 s. In this step, the two photoresists were patterned with the layout of the conductive electrodes where the UV exposure was performed.

Metal deposition was performed in an e-beam evaporator (Neva 005). First, a 50 nm layer of titanium was used to promote adhesion, then a 500 nm thin gold film was deposited. The lift-off process was completed in acetone to allow the photoresist to dissolve while leaving behind the metal pattern. This process was used for electrode gold patterning on the surface of Si wafer. Wafer cleaning was performed in a solvent mixture (acetone, isopropyl, and deionized water) at boiling temperature.

To obtain a passivation of the gold conductive lines, the SU-8 2015 was spin coated at 3000 rpm for 30 s and then pre-baked at 65 °C for 1 min and 95 °C for 5 min. The SU-8 resist was exposed using a photolithographic mask using the MA6 mask aligner for 8 s, followed by post-baking on a hot plate at 65 °C for 1 min and then at 95 °C for 6 min, and developed in mr-DEV-600 solution for 2 min. The wafer was then washed in isopropyl alcohol to stop the action of the developer. To guarantee that the SU-8 passivation layer properties did not modify, the wafer was hard baked on a hot plate at 180 °C for 10 min. The designed microelectrodes on the Si wafer were drawn and cut individually.

The geometry of the interdigitated microelectrodes was tailored in accordance with the cells under study. A castellated architecture was selected for ensuring the development of higher gradient field regions. Each interdigitated microelectrode array had 16 fingers with a length of 2560 µm, the gap between the fingers and the intercastellations had a dimension of 40 µm.

2.2. Cell Culture and Sample Preparation

2.2.1. Cell Lines

Human colon adenocarcinoma cell line SW-403 (Cat. No. 87071008), human leukemic T cell line Jurkat E6.1 (Cat No. ECACC 88042803), and human monocyte-like THP-1 cells (Cat. No. 880881201) were purchased from the European Collection of Authenticated Cell Cultures (ECACC) and cultured in RPMI-1640 (Bio Whittaker Lonza, Verviers, Belgium), supplemented with 10% fetal bovine serum (FBS, Euroclone, Milan, Italy) and 100 IU/mL penicillin + 100 µg/mL streptomycin (Lonza, Basel, Switzerland) (complete culture medium). Cell lines were incubated at 37 °C in an atmosphere supplemented with 5% CO_2, in 75 cm^2 flasks. The adherent cell line, SW-403, was cultured until 85% confluence, then washed with phosphate buffered saline (PBS, Merck, Darmstadt, Germany), and detached using 0.05% trypsin-EDTA solution (Thermo Fischer Scientific, Waltham, MA, USA). The cells were then suspended in a complete culture medium, washed by centrifugation at 200× g for 10 min and then resuspended in a fresh complete culture medium.

Non-adherent cell lines (THP-1 and Jurkat) were simply collected and centrifuged in the previously mentioned conditions.

2.2.2. Isolation and Culture of Primary Tumor Cells

Tumor samples (T1 and T2) were collected from two colorectal cancer patients after written informed consent from each subject and approval from the Ethics Committees of Bucharest Emergency University Hospital and processed as previously described [24]. Briefly, the tumor specimens were excised carefully and aseptically during surgery and transferred to 50 mL tubes with PBS supplemented with antibiotics (100 IU/mL penicillin, 100 μg/mL streptomycin, 1 mg/mL gentamicin, and 0.5 mg/mL vancomycin). Tumor tissues were then transferred to Petri dishes and rinsed with fresh AIM-V containing AlbuMAX® supplement (bovine serum albumin) medium (Thermo Fischer Scientific, Waltham, MA, USA). After resection of the fatty and connective tissues and the necrotic areas, the tumor specimens were minced with sterile scalpels and scissors into small pieces (0.5–1 mm^3) and cultured in AIM-V AlbuMAX supplemented with antibiotics (100 IU/mL penicillin, 100 μg/mL streptomycin, 20 μg/mL gentamicin (Merck, Darmstadt, Germany), and 6 μg/mL vancomycin (Merck, Darmstadt, Germany)) and amphotericin B (5 μg/mL) (Merck, Darmstadt, Germany). The primary tumor cells were maintained in culture continuously for more than 12 months. The cancer cells grew as floating spheroids/aggregates, firmly/loosely adherent spheroids, or as both adherent and floating spheroids/aggregates. Subsequent passages were performed every two or four weeks. To obtain single cells, spheroids/aggregates were dissociated by enzymatic digestion using Accumax-Cell aggregate dissociation medium (Thermo Fischer Scientific, Waltham, MA, USA).

2.2.3. Human Peripheral Blood Mononuclear Cells

Human blood was obtained from a healthy donor (lab worker) after obtaining informed consent and ethical approval. Peripheral blood mononuclear cells (PBMCs) were isolated by using Ficoll-Hypaque (1.077 g/mL density, (Merck, Darmstadt, Germany)) and resuspended in RPMI medium.

2.2.4. Suspension Medium

The low conductivity suspension medium (250 mM sucrose, 13 mS/m conductivity) was chosen based on viability data in preliminary experiments and was prepared by dissolving sucrose (Merck, Darmstadt, Germany) in distilled water and adjusting the pH to 7.4. The osmolarity was measured with a VAPRO Vapor Pressure Osmometer Model 5600 and was 250 mmol/kg. The conductivity measurement was performed with a ZetaSizer Nano-2S. The baseline value was 0.5 mS/m. To increase the conductivity to 13 mS/m, a 250 mM HEPES (Merck, Darmstadt, Germany) solution was used. The pH and conductivity values remained stable for at least one week when stored at 4 °C.

2.2.5. Sample Preparation and Viability Assay

Cancer cell lines, primary tumor cells, and normal PBMC were washed and resuspended in a low conductivity suspension medium (2×10^6 cells/mL). Their viability was evaluated before and after DEP measurements by staining the cells with acridine orange (Merck, Darmstadt, Germany) and propidium iodide (Merck, Darmstadt, Germany) and examining them in a fluorescence microscope (Nikon TE2000) at 100X magnification. Cells that were fluorescing green were scored as viable while cells that were fluorescing orange, either fully or partially, were scored as nonviable.

2.3. Experimental Set-Up and Equipment

The experimental activity in this study involved trapping the cells via dielectrophoresis to determine the crossover frequency by observing the cells' motion and characterization of the trapped cells via electrical impedance spectroscopy. Figure 1 depicts a schematic diagram of the proposed experimental structure. The set-up operation procedure involved two main steps: (1) trapping the cells at the microelectrode level via DEP and (2) identification of the cell's type by measuring its impedance characteristics. The test section of the microchip consisted of the electrode substrate on the bottom and a glass cover on the top for observation of the cells. For the DEP experiments, a Keysight 33521A Function/Arbitrary Waveform Generator was employed to generate a sinusoidal AC electric field. The cell's distribution at microelectrode level was monitored and recorded using an improvised optical setup consisting of a Nikon Plan Fluor 10x/0.30 microscope objective with a mounted CCD Nikon Digital Sight DS-Qi1Mc camera connected to a computer that was running NIS-Elements AR 3.0 SP 1 (Build 455) software. Electrical impedance spectroscopy measurements were performed using a Novocontrol Broadband Dielectric Spectrometer (Alpha-A High-Performance Frequency Analyzer). The electrodes were connected to the analyser and generator by using a Micrux drop-cell connector. The impedance experimental data were fitted with the software EIS Spectrum Analyser 1.0 program [25].

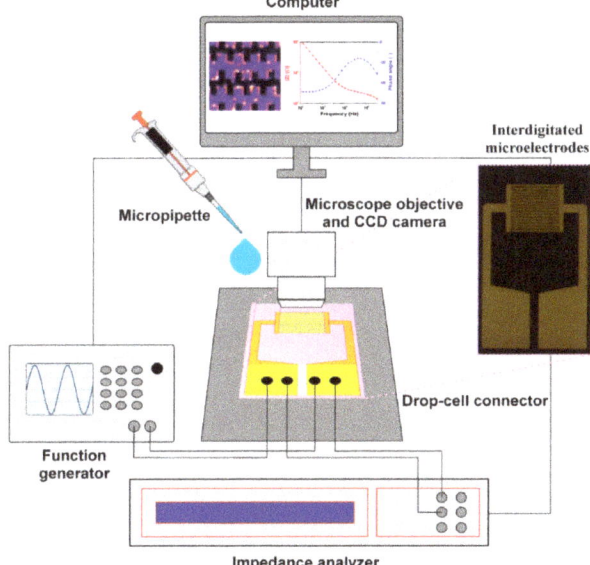

Figure 1. Experimental set-up.

3. Results and Discussion

3.1. DEP-Based Cells Manipulation and Electrical Impedance Spectroscopy Measurement

In the presence of an inhomogeneous electric field gradient, the biological cells may be displaced towards the electric field maxima (positive DEP) or towards the electric field minima (negative DEP) depending on the dielectric properties of the specific cell and on the properties of the suspending media. In a first experiment, normal (PBMC) and tumor cells (SW-403 cell line and primary tumor cells T1) were subjected to a sinusoidal excitation voltage (9 V peak-to-peak magnitude and a frequency of 1 MHz) that was applied to the electrodes, for approximately 5 min to concentrate the cells on the electrodes. Under the effect of p-DEP, after few seconds (≈4 s) the cells concentrated at the electrode surface (see the Supplementary Material). Figure 2 depicts the microscopic images of cell samples before and after (5 min) DEP manipulation. It is visible that under these experimental conditions the majority of the cells of both normal and tumor cell populations were displaced towards the highest electric field regions. Moreover, in some regions the cells followed electric field lines between adjacent microelectrodes due to their high interfacial polarization, creating "cells' bridges". Before and after the cells were trapped at the level of microelectrodes, the impedance measurements were carried out to differentiate the normal cells from cancer ones.

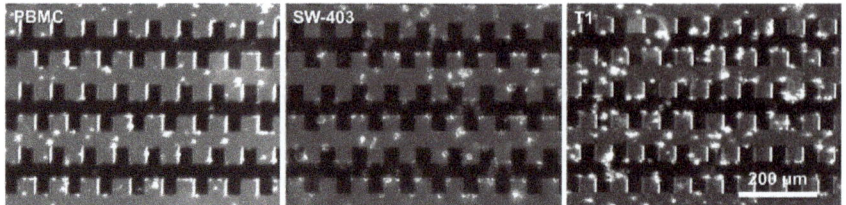

Figure 2. Microscopy images of a cell samples (PBMC, SW-403 and T1) distributions after DEP manipulation.

Next, the impedance measurements were performed on normal cells (PBMC and THP-1-monocyte cell line), two tumor cell lines (an adherent adenocarcinoma cell line (SW-403) and Jurkat, a non-adherent T cell line), and the primary tumor cells isolated from two colon cancer patients (T1 and T2).

The impedance measurements of the un-trapped and trapped living cells were performed in the frequency range from 0.1 to 300 kHz at an operating voltage of 100 mV. This frequency range was selected to monitor the evolution of the electrical properties of each cell type in the α and β dispersion regions. The frequency range was selected for exploring the effect of ionic diffusion and interfacial polarization of biological membrane systems [26]. Figure 3 depicts the measured electrical impedance spectra (amplitude Z, phase angle θ, and Nyquist plots) of the three cell types (cancer cell lines, primary tumor cells, and normal PBMCs) before and after the DEP concentration at the electrode level. The impedance magnitude of the suspension medium (in the absence of cells) decreased when the frequency increased. The transition from capacitive behaviour, which dominates at lower frequencies, to the resistive behaviour, that prevails at higher frequencies, was highlighted. Generally, adding the cells to the suspension medium lead to an augmentation of the total impedance as compared to the medium alone (Figure 3a).

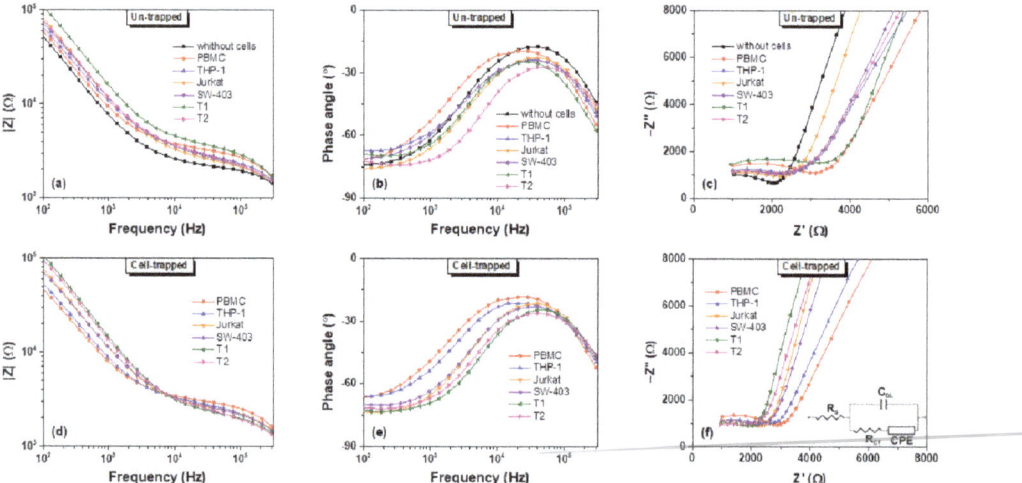

Figure 3. Electrical impedance spectroscopy responses (amplitude Z (**a**,**d**), phase angle (**b**,**e**), and Nyquist plots (**c**,**f**)) of different cell types suspended in buffered sucrose solution, before and after cells trapping.

However, before DEP manipulation, when the cells are suspended within the entire volume of the suspension solution, no significant differences among the different types of cells could be noticed. The presence of each cell sample uniquely changed the impedance response of the suspension medium. By contrast, in the presence of DEP forces that were applied for 5 min, a different feature below 10 kHz was noticed on the phase angle and Nyquist (Figure 3e,f) characteristics. The THP-1 cell line is regarded as a model for primary monocytes (i.e., monocytes from human peripheral blood), which explains their similarity in terms of dielectric responses. As a consequence of cell migration onto the electrode surface, the local ionic environment at the electrode/electrolyte interface was affected due to high insulating of the cell membranes; cell trapping lead to a decrease of electrode surface area and therefore an increase of the interface impedance. At low frequencies, the current was forced to flow between the insulating cell membranes, while at higher frequencies the current penetrated the cell membranes and flowed through the intracellular and extracellular fluid [27,28]. Therefore, differences noticed at frequencies below 10 kHz between normal cells (PBMC and THP-1) and cancer cells (Jurkat, SW-403, T1, and T2) may be attributed to the surface morphological features of the cell membranes and to the electrode surface area which is covered with cells (i.e., the cells radii). At higher frequencies, the spectrum of the total impedance is presumably influenced by the suspension medium, reaching almost the same value for all cell types, as can be seen in Figure 3d.

The functionality and reproducibility of the proposed method was evaluated from the EIS responses of trapped cells at different DEP operating voltages with five independent interdigitated microelectrodes that were fabricated by a similar procedure. Based on the impedance magnitude and the phase angle frequency dependences that are illustrated in Figure S1 (Supplementary Material), no significant differences in the EIS responses were observed in all five individual microelectrodes that were employed for trapping T2 cancer cells at 9 Vpp and 1 MHz. Table S1 depicts the average values of the impedance magnitude and phase angle, the standard deviation (SD), and the relative standard deviation (RSD) that was calculated between the electrodes at three different frequencies (10^3, 10^4, and 10^5 Hz). The reproducibility of our manufactured interdigitated microelectrodes was found to be good with a %RSD yield in the range of 2.24 to 6.44%. Furthermore, method reproducibility was evaluated by analyzing the influence of the DEP operating voltages (3, 6, 9, and 12 Vpp) on the electrical characterization of PBMC, THP-1, and T2 cells (Figure S2). Even if the DEP voltage was changed, the impedance spectra were similar with minor variations at low frequencies in the case of PBMC and T2 cells obtained from donors due to their heterogeneity.

To understand which specific characteristics influenced the different features of the normal and cancer cells, the DEP crossover frequency experiment and the electrical equivalent circuit model were used.

The DEP crossover frequency (f_{co}) is the characteristic frequency at which the polarity of the dielectrophoretic force changes and cells experience zero DEP force. By observing the motion of cells at the electrode edges when the frequency that is applied is slowly swept, the f_{co} of each cell type can be ascertained [3,29]. In our dielectrophoretic crossover frequency experiment, the microchip was powered by an AC voltage with 12 Vpp of variable frequency at the level of two adjacent microelectrodes. It should be mentioned that the DEP operating voltage was not affecting the impedance spectra (please see the Supplementary information file, Figure S2) when the experiments were running during the same period of time, however, for the crossover frequency experiments we choose 12 Vpp voltage as the displacement of the cells is more visible. The voltage frequency was sequentially increased from 10 kHz up to 1 MHz and the cell displacements induced by the DEP force were examined with a microscope. The crossover frequency at which the cell exercised no DEP movement was recorded. Within individual experiments, at least 10 frequencies were determined for each cell type and all measurements were performed at room temperature.

Figure 4 depicts the experimentally determined crossover frequencies for various human cancer cells, including the primary tumor cells (T1 and T2) that were collected from two colorectal cancer patients, a colon adenocarcinoma cell line (SW-403), a human leukemic T cell line (Jurkat), a human monocyte-like cell line (THP-1), and peripheral blood mononuclear cells (PBMCs) from a healthy subject, that were all suspended in medium with a conductivity of 13 mS/m. As expected, THP-1, Jurkat, and SW-403 cancer cell lines exhibited distinct behaviour, characterized by lower average crossover frequencies (57.4 ± 2.5 kHz, 31.6 ± 1.7 kHz, and 28.2 ± 1.4 kHz, respectively) in comparison to PBMCs (106.2 ± 5.4 kHz), which allowed discrimination of each type of cell. Moreover, the primary tumor cells (T1 and T2) presented characteristic crossover frequencies within the same domain of frequency as also observed for the cancer cell lines. According to the literature, these different DEP frequency responses of cancer and normal blood cells may be explained and expressed by Gascoyne and Shim [7] in terms of reciprocal cell "dielectric phenotype" $1/R\phi$, where ϕ represents the membrane folding factor (the ratio of actual membrane area to that of the idealized smooth shell) and R is the cell radius. Many studies have reported that cancer cells have a larger folding factor and radii than both blood cells and normal cells of comparable origin [4,5,8,30–34]. A plausible explanation could be related to an increase in the membrane cholesterol or the membrane lipid rafts in cancer cells [35,36].

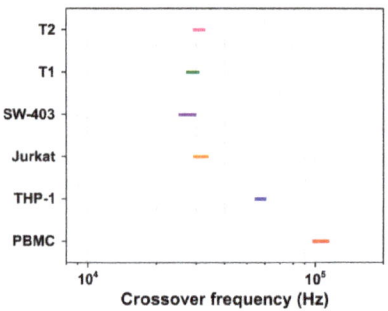

Figure 4. The DEP crossover frequency for the different types of cancer cells and healthy peripheral blood mononuclear cells.

Due to the notorious heterogeneity of cancer cells, especially of the primary tumor cells, it was difficult to estimate R and ϕ parameters for each cell type. Thus, the following discussions are based on the reciprocal dielectric phenotype which is proportional to the DEP crossover frequency:

$$f_{co} \approx \frac{1}{\sqrt{2}\pi C_0}\left(\frac{\sigma_s}{R\phi}\right) \quad (1)$$

where σ_s is the conductivity of the suspending medium and $C_0 = 9$ mF/m² [37] represents the specific capacitance of the smooth cell plasma membrane. The calculated reciprocal cell dielectric phenotype (Table S2) demonstrated notable differences between the cancer and normal peripheral blood mononuclear cells, highlighting the fact that the dielectric response of each cell type is influenced significantly by its morphological characteristics (i.e., its size and shape).

3.2. Interpretation of Measured Impedance Data by Equivalent Circuit

To explain the electrical impedance characteristics of the cell-covered electrode, an electrical equivalent circuit model was used. The experimental impedance spectra (Nyquist plots) were analysed in accordance to Randles equivalent circuit model [38] (Figure 3f, inset). The electrolyte's resistance, R_S, represents the suspension medium in series with a parallel group of double layer capacitance C_{DL} necessary for the charging of the electrode/electrolyte interface and charge transport element, that is represented by a series group of charge transport resistance R_{CT} and a constant phase element CPE.

Since cell membranes, ideally modelled as capacitors, include a lipid bilayer, surface roughness, and integrated ion channels that resemble a porous surface contact, the capacitance was modulated by the charge transfer and differs from the capacitance of an ideal capacitor (i.e., frequency-dependent) [39,40]. Therefore, the R_{CT} and CPE series group is describing the transport phenomena near the electrodes [40–42] (i.e., the charge transport through the electrode/electrolyte interface including the cells membrane capacitance (electrode–cells–suspension medium assembly)). Under these considerations, the total measured impedance Z of the system can be expressed as:

$$Z = R_S + \frac{Z_{DL}(R_{CT} + Z_{CPE})}{Z_{DL} + R_{CT} + Z_{CPE}} \qquad (2)$$

where $Z_{DL} = \frac{1}{j\omega C_{DL}}$ and $Z_{CPE} = \frac{1}{Q(j\omega)^n}$ where Q is a measure of the magnitude of Z_{CPE}, ω is the angular frequency, and n is a constant ($0 \leq n \leq 1$).

By fitting the impedance measurements after DEP trapping, as we expected, the extracted resistance of the solutions and double layer capacitances were similar for all of the types of cells involved in our experiment, with an average value of $R_S = 267 \pm 7.8\ \Omega$ and $C_{DL} = 342 \pm 16.9$ pF. As shown in Figure 5a, the extracted charge transport resistances R_{CT} of tumor cells T1 and T2 were approximately equal but their values were lower than the ones of normal PBMC cells. It was noticeable that the value of parameter R_{CT} for PBMC cells was higher in comparison to values of the cancerous lines that were involved in the study even if, in the case of THP-1, the difference was not considerable. The extracted magnitudes Q and n constant of Z_{CPE} for the cancer cells were in the range of $4.0 \times 10^{-8} \pm 1.16 \times 10^{-9}$–$6.5 \times 10^{-8} \pm 2.07 \times 10^{-9}$ s^n/Ω and 0.830–0.855, respectively, while those for the normal cells were in the range of $1.7 \times 10^{-7} \pm 8.78 \times 10^{-9}$–$1.9 \times 10^{-7} \pm 8.35 \times 10^{-9}$ s^n/Ω and 0.762–0.763, respectively (Figure 5b,c). Moreover, the values of Q and n of PBMC cells were very different in comparison to the values of the same parameters of Jurkat, SW-403, T1, and T2 tumor cells but close to the values for THP-1 cells. The less evident difference between the PBMC and THP-1 may be attributed to the fact that, as stated in Section 3.1, the THP-1 cell line is regarded as a model for primary monocytes. Moreover, the fact that under p-DEP, cells migrated towards the electrode interface, as is visible in Figure 2, so all charge transport phenomena at this interface is mediated and altered by these cells. Thus, the Randles circuit transport elements, R_{CT} and CPE, were influenced by the cells' size and morphological characteristics (i.e., their dielectric phenotype), especially the cell membrane features since they facilitated all of the charge transport to and from the extracellular medium.

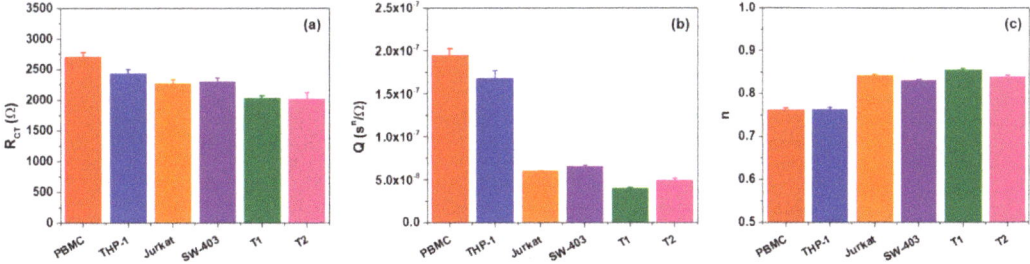

Figure 5. The electrical fitting parameters (R_{CT} (**a**), Q (**b**), and n (**c**)) in the equivalent circuit model for studied cells. Error bars indicate the values of the relative estimated errors of the calculated parameters.

4. Conclusions

We reported a study proposing the combined utilization of EIS and DEP for enabling the rapid and label-free differentiation of various cancer cells from normal ones. The method's successful exploitation was based on the correlation of impedance characteristics of the cells with their biophysical phenotype. Experiments were performed using interdigitated microelectrodes and included three cancerous cell lines, two types of primary tumor cells, and normal blood cells. Crossover frequencies that were determined during the application of DEP forces between different types of cells achieved reasonably different values. The impedance spectra after DEP trapping demonstrated that an electrical signature may be a future solution in differentiating cancer cells from normal cells. Moreover, the Randles equivalent circuit model highlighted differences between a series group of charge transport resistance and constant phase elements for cancerous and normal cells fact which were assigned to a dielectric phenotype. Through its high capacity for discrimination, the proposed method could be a valuable approach for the detection of circulating tumor cells (CTCs).

Supplementary Materials: The following are available online at https://www.mdpi.com/article/10.3390/bios11100401/s1, Figure S1: The microelectrodes reproducibility: the impedance magnitudes and phase angles of the trapped T2 cancer cells at 9 Vpp, 1 MHz obtained for the five independent interdigitated microelectrodes; Table S1: The summary of the reproducibility tests by EIS technique performed at the level of five independent interdigitated microelectrodes; Figure S2: The impedance magnitudes of trapped PBMC, THP-1, and T2 cells at 3, 6, 9, and 12 Vpp; Table S1: The reciprocal cell dielectric phenotype of cancer and normal blood cells; video of a THP-1 cells concentrated at the electrode surface.

Author Contributions: I.T.: Conceptualization, Methodology, Validation, Investigation, Formal analysis, Writing—Original draft preparation; I.C.: Supervision, Methodology, Investigation; T.G.S.: Methodology and Investigation; C.T.: Conceptualization, Methodology, Software; A.S.: Funding acquisition, Writing—Reviewing and Editing; V.V.: Methodology, Investigation, Data curing, Software, Writing—Original draft preparation; M.A.: Methodology, Conceptualization, Project administration. B.T.: Conceptualization, Methodology, Investigation. M.A.O.: Project administration, Funding acquisition, Supervision, Conceptualization, Methodology, Investigation, Writing—Reviewing and Editing. All authors have read and agreed to the published version of the manuscript.

Funding: This work was supported by a publications grant from the Technical University of Iasi (TUIASI), project number no. GI/R 15 DRD/2021.

Institutional Review Board Statement: The study was conducted according to the guidelines of the Declaration of Helsinki, and approved by the Ethics Committees of Bucharest Emergency University Hospital (51481, 30 October 2018).

Informed Consent Statement: Informed consent was obtained from all subjects that were involved in the study.

Data Availability Statement: The data that support the findings of this study are available on request from the corresponding author.

Acknowledgments: This work was supported by a grant of the Romanian Ministry of Research and Innovation, CCCDI-UEFISCDI, Project Number PN-III-P1-1.2-PCCDI-2017-0214/3PCCDI, within PNCDI III.

Conflicts of Interest: The authors declare no conflict of interest.

References

1. Guido, I.; Xiong, C.; Jaeger, M.S.; Duschl, C. Microfluidic system for cell mechanics analysis through dielectrophoresis. *Microelectron. Eng.* **2012**, *97*, 379–382. [CrossRef]
2. Alshareef, M.; Metrakos, N.; Juarez Perez, E.; Azer, F.; Yang, F.; Yang, X.; Wang, G. Separation of tumor cells with dielectrophoresis-based microfluidic chip. *Biomicrofluidics* **2013**, *7*, 011803. [CrossRef] [PubMed]
3. Huang, C.; Liu, C.; Minne, B.; Ramirez Hernandez, J.E.; Stakenborg, T.; Lagae, L. Dielectrophoretic discrimination of cancer cells on a microchip. *Appl. Phys. Lett.* **2014**, *105*, 143702. [CrossRef]

4. Broche, L.M.; Bhadal, N.; Lewis, M.P.; Porter, S.; Hughes, M.P.; Labeed, F.H. Early detection of oral cancer-Is dielectrophoresis the answer? *Oral Oncol.* **2007**, *43*, 199–203. [CrossRef]
5. Mulhall, H.J.; Labeed, F.H.; Kazmi, B.; Costea, D.E.; Hughes, M.P.; Lewis, M.P. Cancer, pre-cancer and normal oral cells distinguished by dielectrophoresis. *Anal. Bioanal. Chem.* **2011**, *401*, 2455–2463. [CrossRef]
6. Labeed, F.H.; Coley, H.M.; Hughes, M.P. Differences in the biophysical properties of membrane and cytoplasm of apoptotic cells revealed using dielectrophoresis. *Biochim. Biophys. Acta-Gen. Subj.* **2006**, *1760*, 922–929. [CrossRef]
7. Gascoyne, P.R.C.; Shim, S. Isolation of circulating tumor cells by dielectrophoresis. *Cancers* **2014**, *6*, 545–579. [CrossRef]
8. Gascoyne, P.R.C.; Shim, S.; Noshari, J.; Becker, F.F.; Stemke-Hale, K. Correlations between the dielectric properties and exterior morphology of cells revealed by dielectrophoretic field-flow fractionation. *Electrophoresis* **2013**, *34*, 1042–1050. [CrossRef]
9. Gupta, V.; Jafferji, I.; Garza, M.; Melnikova, V.O.; Hasegawa, D.K.; Gupta, V.; Jafferji, I.; Garza, M.; Melnikova, V.O. ApoStream^{TM}, a new dielectrophoretic device for antibody independent isolation and recovery of viable cancer cells from blood. *Biomicrofluidics* **2012**, *6*, 024133. [CrossRef]
10. Nguyen, N.V.; Jen, C.P. Impedance detection integrated with dielectrophoresis enrichment platform for lung circulating tumor cells in a microfluidic channel. *Biosens. Bioelectron.* **2018**, *121*, 10–18. [CrossRef]
11. Zhao, Y.; Zhao, X.T.; Chen, D.Y.; Luo, Y.N.; Jiang, M.; Wei, C.; Long, R.; Yue, W.T.; Wang, J.B.; Chen, J. Tumor cell characterization and classification based on cellular specific membrane capacitance and cytoplasm conductivity. *Biosens. Bioelectron.* **2014**, *57*, 245–253. [CrossRef] [PubMed]
12. Qian, C.; Huang, H.; Chen, L.; Li, X.; Ge, Z.; Chen, T.; Yang, Z.; Sun, L. Dielectrophoresis for bioparticle manipulation. *Int. J. Mol. Sci.* **2014**, *15*, 18281–18309. [CrossRef] [PubMed]
13. Turcan, I.; Olariu, M.A. Dielectrophoretic Manipulation of Cancer Cells and Their Electrical Characterization. *ACS Comb. Sci.* **2020**, *22*, 554–578. [CrossRef]
14. Marusyk, A.; Polyak, K. Tumor heterogeneity: Causes and consequences. *Biochim. Biophys. Acta-Rev. Cancer* **2010**, *1805*, 105–117. [CrossRef] [PubMed]
15. Marjanovic, N.D.; Weinberg, R.A.; Chaffer, C.L. Cell Plasticity and Heterogeneity in Cancer. *Clin. Chem.* **2013**, *59*, 168–179. [CrossRef] [PubMed]
16. Marusyk, A.; Almendro, V.; Polyak, K. Intra-tumour heterogeneity: A looking glass for cancer? *Nat. Rev. Cancer* **2012**, *12*, 323–334. [CrossRef]
17. Crowell, L.L.; Yakisich, J.S.; Aufderheide, B.; Adams, T.N.G. Electrical impedance spectroscopy for monitoring chemoresistance of cancer cells. *Micromachines* **2020**, *11*, 832. [CrossRef]
18. Qiao, G.; Duan, W.; Chatwin, C.; Sinclair, A.; Wang, W. Electrical properties of breast cancer cells from impedance measurement of cell suspensions. *J. Phys. Conf. Ser.* **2010**, *224*, 2–6. [CrossRef]
19. Huerta-Nuñez, L.F.E.; Gutierrez-Iglesias, G.; Martinez-Cuazitl, A.; Mata-Miranda, M.M.; Alvarez-Jiménez, V.D.; Sánchez-Monroy, V.; Golberg, A.; González-Díaz, C.A. A biosensor capable of identifying low quantities of breast cancer cells by electrical impedance spectroscopy. *Sci. Rep.* **2019**, *9*, 1–12. [CrossRef]
20. Giana, F.E.; Bonetto, F.J.; Bellotti, M.I. Assay based on electrical impedance spectroscopy to discriminate between normal and cancerous mammalian cells. *Phys. Rev. E* **2018**, *97*, 1–10. [CrossRef]
21. Han, A.; Yang, L.; Frazier, A.B. Quantification of the heterogeneity in breast cancer cell lines using whole-cell impedance spectroscopy. *Clin. Cancer Res.* **2007**, *13*, 139–143. [CrossRef] [PubMed]
22. Al Ahmad, M.; Al Natour, Z.; Mustafa, F.; Rizvi, T.A. Electrical Characterization of Normal and Cancer Cells. *IEEE Access* **2018**, *6*, 25979–25986. [CrossRef]
23. Zhang, F.; Jin, T.; Hu, Q.; He, P. Distinguishing skin cancer cells and normal cells using electrical impedance spectroscopy. *J. Electroanal. Chem.* **2018**, *823*, 531–536. [CrossRef]
24. Caras, I.; Tucureanu, C.; Lerescu, L.; Pitica, R.; Melinceanu, L.; Neagu, S.; Salageanu, A. Influence of tumor cell culture supernatants on macrophage functional polarization: In vitro models of macrophage-tumor environment interaction. *Tumori* **2011**, *97*, 647–654. [CrossRef] [PubMed]
25. Bondarenko, A.S.; Ragoisha, G.A. Inverse problem in potentiodynamic electrochemical impedance spectroscopy. In *Progress in Chemometrics Research*; Pomerantsev, A.L., Ed.; Nova Science: New York, NY, USA, 2005; pp. 89–102.
26. Schwan, H.P. Electrical Properties of Tissue and Cell Suspensions *. In *Advances in Biological and Medical Physics*; Academic Press. Inc.: Cambridge, MA, USA, 1957; Volume 5, pp. 147–209.
27. Amini, M.; Hisdal, J.; Kalvøy, H. Applications of bioimpedance measurement techniques in tissue engineering. *J. Electr. Bioimpedance* **2018**, *9*, 142–158. [CrossRef]
28. Stolwijk, J.A.; Wegener, J. Impedance-Based Assays Along the Life Span of Adherent Mammalian Cells In Vitro: From Initial Adhesion to Cell Death. In *Label-Free Monitoring of Cells in vitro. Bioanalytical Reviews*; Springer: Cham, Switzerland, 2019; Volume 2, pp. 1–75.
29. Vykoukal, D.M.; Gascoyne, P.R.C.; Vykoukal, J. Dielectric characterization of complete mononuclear and polymorphonuclear blood cell subpopulations for label-free discrimination. *Integr. Biol.* **2009**, *1*, 477–484. [CrossRef]
30. Shim, S.; Stemke-Hale, K.; Noshari, J.; Becker, F.F.; Gascoyne, P.R.C. Dielectrophoresis has broad applicability to marker-free isolation of tumor cells from blood by microfluidic systems. *Biomicrofluidics* **2013**, *7*, 011808. [CrossRef]

31. An, J.; Lee, J.; Lee, S.H.; Park, J.; Kim, B. Separation of malignant human breast cancer epithelial cells from healthy epithelial cells using an advanced dielectrophoresis-activated cell sorter (DACS). *Anal. Bioanal. Chem.* **2009**, *394*, 801–809. [CrossRef]
32. Wu, L.; Yung, L.-Y.L.; Lim, K.-M. Dielectrophoretic capture voltage spectrum for measurement of dielectric properties and separation of cancer cells. *Biomicrofluidics* **2012**, *6*, 14113–1411310. [CrossRef]
33. Liu, N.; Lin, Y.; Peng, Y.; Xin, L.; Yue, T.; Liu, Y.; Ru, C.; Xie, S.; Dong, L.; Pu, H.; et al. Automated Parallel Electrical Characterization of Cells Using Optically-Induced Dielectrophoresis. *IEEE Trans. Autom. Sci. Eng.* **2020**, *17*, 1084–1092. [CrossRef]
34. Mohamed, R.; Razak, M.A.A.; Kadri, N.A. Determination of electrophysiological properties of human monocytes and THP-1 cells by dielectrophoresis. *Biomed. Res. Ther.* **2019**, *6*, 3040–3052. [CrossRef]
35. Li, Y.C.; Park, M.J.; Ye, S.K.; Kim, C.W.; Kim, Y.N. Elevated levels of cholesterol-rich lipid rafts in cancer cells are correlated with apoptosis sensitivity induced by cholesterol-depleting agents. *Am. J. Pathol.* **2006**, *168*, 1107–1118. [CrossRef]
36. Hanahan, D.; Weinberg, R.A. Hallmarks of cancer: The next generation. *Cell* **2011**, *144*, 646–674. [CrossRef] [PubMed]
37. Pethig, R.; Kells, D.B. *The Passive Electrical Properties of Biological Systems: Their Significance in Physiology, Biophysics and Biotechnology*; IOP Publishing Ltd.: Bristol, UK, 1987; Volume 32.
38. Randles, J.E.B. Kinetics of rapid electrode reactions. *Faraday Discuss.* **1947**, *1*, 11–19. [CrossRef]
39. Mesa, F.; Paez-Sierra, B.A.; Romero, A.; Botero, P.; Ramírez-Clavijo, S. Assisted laser impedance spectroscopy to probe breast cancer cells. *J. Phys. D Appl. Phys.* **2021**, *54*, 075401. [CrossRef]
40. Mojena-Medina, D.; Hubl, M.; Bäuscher, M.; Jorcano, L.; Ngo, H.; Acedo, P. Real-Time Impedance Monitoring of Epithelial Cultures with Inkjet-Printed Interdigitated-Electrode Sensors. *Sensors* **2020**, *20*, 5711. [CrossRef]
41. Kadan-Jamal, K.; Sophocleous, M.; Jog, A.; Desagani, D.; Teig-Sussholz, O.; Georgiou, J.; Avni, A.; Shacham-Diamand, Y. Electrical Impedance Spectroscopy of plant cells in aqueous biological buffer solutions and their modelling using a unified electrical equivalent circuit over a wide frequency range: 4Hz to 20 GHz. *Biosens. Bioelectron.* **2020**, *168*, 112485. [CrossRef]
42. Pradhan, R.; Mitra, A.; Das, S. Impedimetric characterization of human blood using three-electrode based ECIS devices. *J. Electr. Bioimpedance* **2012**, *3*, 12–19. [CrossRef]

Article

Design and Analysis of a Single System of Impedimetric Biosensors for the Detection of Mosquito-Borne Viruses

Fahmida Nasrin [1], Kenta Tsuruga [2], Doddy Irawan Setyo Utomo [3], Ankan Dutta Chowdhury [1] and Enoch Y. Park [1,2,3,*]

1. Laboratory of Biotechnology, Research Institute of Green Science and Technology, Shizuoka University, 836 Ohya, Suruga-ku, Shizuoka 422-8529, Japan; fahmida.nasrin.17@shizuoka.ac.jp (F.N.); ankan.dutta.chowdhury@shizuoka.ac.jp (A.D.C.)
2. Laboratory of Biotechnology, Department of Agriculture, Graduate School of Integrated Science and Technology, Shizuoka University, 836 Ohya, Suruga-ku, Shizuoka 422-8529, Japan; tsuruga.kenta.17@shizuoka.ac.jp
3. Laboratory of Biotechnology, Graduate School of Science and Technology, Shizuoka University, 836 Ohya, Suruga-ku, Shizuoka 422-8529, Japan; doddy.irawan.setyo.utomo.16@shizuoka.ac.jp
* Correspondence: park.enoch@shizuoka.ac.jp

Abstract: The treatment for mosquito-borne viral diseases such as dengue virus (DENV), zika virus (ZIKV), and chikungunya virus (CHIKV) has become difficult due to delayed diagnosis processes. In addition, sharing the same transmission media and similar symptoms at the early stage of infection of these diseases has become more critical for early diagnosis. To overcome this, a common platform that can identify the virus with high sensitivity and selectivity, even for the different serotypes, is in high demand. In this study, we have attempted an electrochemical impedimetric method to detect the ZIKV, DENV, and CHIKV using their corresponding antibody-conjugated sensor electrodes. The significance of this method is emphasized on the fabrication of a common matrix of gold–polyaniline and sulfur, nitrogen-doped graphene quantum dot nanocomposites (Au-PAni-N,S-GQDs), which have a strong impedimetric response based only on the conjugated antibody, resulting in minimum cross-reactivity for the detection of various mosquito-borne viruses, separately. As a result, four serotypes of DENV and ZIKV, and CHIKV have been detected successfully with an LOD of femtogram mL^{-1}.

Keywords: dengue virus; dengue serotype; mosquito-borne viral disease; virus detection; electrochemical impedance spectroscopy

1. Introduction

The severity of mosquito-borne diseases is a global threat caused by protozoa, viruses, or parasites, resulting in nearly 700 million illnesses and over one million deaths each year [1]. The annual epidemic for protozoa-caused malaria has been a deadly problem in tropical regions since the nineteenth century. However, in recent decades, the outbreaks of diseases such as dengue, zika, and chikungunya viruses are widely epidemic throughout the summer and rainy season [2–4]. Zika virus (ZIKV), dengue virus (DENV), and chikungunya virus (CHIKV) are vector-borne human viral pathogens, sharing the same vectors of Aedes aegypti or Aedes albopictus [5,6]. The fatality of these viral diseases is significantly high, especially in highly populated regions due to their fast transmission rates and delays in initial diagnosis. Symptomatic diagnosis is more critical because ZIKV, DENV, and CHIKV share the same transmission media and similar clinical manifestations such as fever, myalgia, and headaches at the early stage of infection [7–9]. In 2019, over 5.2 million dengue cases were reported worldwide: most significantly in tropical and sub-tropical regions such as India, Bangladesh, and Brazil [10]. Chikungunya and zika have exhibited equivalent fatality over the past few decades in India, Brazil, Bangladesh,

Indonesia, etc. Meanwhile, most notably has been that imported cases of ZIKV infection from South America and Oceania were reported in some areas of China, where outbreaks have never previously been reported [11,12]. It indicates that the high spreading ability of these viruses can cause a pandemic if not diagnosed at early stages of infection.

Among the various diagnosis methods, virus isolation, enzyme-linked immunosorbent assay, and reverse-transcriptase polymerase chain reaction (RT-PCR) have generally been used for detecting viral infections [13–15]. Although virus isolation in susceptible cell lines is a highly reliable detection method, it is not an appropriate clinical diagnostic assay for the detection of early infection of these viruses [16,17]. Due to the low level of immunoglobulin M in the early stage and the high possibility of cross-reactivity among these viruses and their subtypes, enzyme-linked immunosorbent assays are also an insufficient method for early diagnosis [18]. Despite some development of nanotechnology-based rapid detection methods, the reliability could not perform adequately close to the gold standard RT-PCR methods [19–22]. However, all these methods have some limitations in replacing the expensive but standard methods such as RT-PCR. For the diagnosis of multiple viruses with typical symptoms, these methods are extremely time-consuming.

Following the advancement of nanotechnologies in virus detection, many reports can be found for direct or indirect ZIKV, CHIKV, and DENV detection [8,23]. Among the different methods for nanomaterial-based biosensors, fluorometric and electrochemical detection methods have emerged recently due to their simple techniques, fast responses, and cost-effectivity [24,25]. In terms of sensitivity, electrochemical methods are always preferable; however, applications for closely related virus samples are rarely reported and need to be studied more thoroughly.

Inspired by our few recent reports on electrochemical sensing [20,26–29], in this study we have developed an electrochemical impedimetric biosensor using gold–polyaniline nanocomposites (Au-PAni) and nitrogen, sulfur co-doped graphene quantum dots (N,S-GQDs) as the base matrix [30,31]. EIS is very popular in energy storage, battery, and solid-state electrolyte applications [32–34], although applications in biosensing devices are also emerging. In this study, the nanocomposites of Au-PAni and N,S-GQDs were conjugated together with different antibodies and thereafter applied for their corresponding target viruses by the impedimetric process. Plenty of carboxylic groups on GQDs can covalently be attached with the free amino group of antibodies, which makes the electrode surface stable and specific for detection. In addition, due to minimum interactions of the Au-PAni-N,S-GQD towards biological substances, including the target virus, the sensor's specificity is solely dependent on the antibody–antigen interaction. Therefore, Au-PAni-N,S-GQD nanocomposites have been used in this study to detect different mosquito-transmitted viruses such as CHIKV, ZIKV, and DENV, altering only the antibody on the sensor surface. Successful results with minimal cross-reactivity encouraged us to proceed with a more intense study on DENV serotype detection. Due to proper optimization of the sensor development and blocking of the electrode surface, the sensor showed good responses in the electrochemical impedimetric signal towards the corresponding viruses, even in different serotypes. In all cases, the limit of detection was found to be as low as femtogram mL^{-1} concentration, which confirms the applicability of this sensor for sensitive and rapid detection. The successful results in this study encourage the extension of this research to explore the sensor performance in multiple detection platforms for different mosquito-borne viruses in single-pot measurements in future.

2. Materials and Methods

2.1. Reagents and Biomaterials

Sodium acetate, sulfuric acid, phosphate-buffered saline (PBS) buffer, aniline, toluene, potassium hydroxide (KOH), hydrochloric acid (HCL), methanol, ethanol, citric acid anhydrous, thiourea, and acetone were purchased from Wako Pure Chemical (Osaka, Japan). Bovine serum albumin (BSA), HAuCl$_4$, N-(3-dimethyl aminopropyl)-N'-ethyl carbodiimide hydrochloride (EDC), and N-hydroxysuccinimide (NHS) were purchased from Sigma Aldrich Co., LLC (Saint Louis, MO, USA). Chikungunya virus lysate (strain: Chikungunya virus), Zika virus lysate (strain: Zika virus, ref 1308258v), Mouse anti-Chikungunya virus capsid protein (clone: CA980), and Mouse anti-Zika virus antibody (Clone: ID5-2-H7-G3) were purchased from The Native Antigen Company (Oxfordshire, UK). Bm5 cells were provided by Prof. K. S. Boo (Insect Pathology Laboratory, School of Agricultural Biotechnology, Seoul National University, Seoul, South Korea) and maintained at 27 °C in Sf-900II serum-free medium (Thermo Fisher Scientific K.K., Tokyo, Japan) supplemented with 1% fetal bovine serum and Antibiotic–Antimycotic solution (Thermo Fisher Scientific K.K., Tokyo, Japan).

The monoclonal anti-dengue virus envelope protein of serotype 1 (Clone E29), serotype 2 (Clone 3H5-1), serotype (Clone E1), and serotype 4 (Clone E42) were purchased from Bei Resources (Manassas, VA, USA). For the selectivity test, influenza virus A/H1N1 (New Caledonia/20/99) was purchased from Prospec-Tany Techno Gene Ltd. (Rehovot, Israel), and norovirus-like particle (NoV-LP) was provided by Dr. Tian-Cheng Li (National Institute of Infectious Diseases, Tokyo, Japan).

2.2. Equipment

Transmission electron microscopy (TEM) images of nanocomposites were obtained using a TEM system (JEM-2100F; JEOL, Ltd., Tokyo, Japan) operated at 100 kV. Dynamic light scattering (DLS) measurements were performed using a Zetasizer Nano series (Malvern Inst. Ltd., Malvern, UK). Electrochemical impedance spectroscopy (EIS) and electrochemical cyclic voltammetry (CV) were performed by using an SP-150 (BioLogic Inc., Tokyo, Japan), which consists of a conventional three-electrode cell containing platinum wire. Saturated Ag/AgCl was used as an electrolyzer (EC frontier, Tokyo, Japan).

2.3. Preparation of the AuNP-PAni Nanocomposite

AuNP-PAni nanocomposites were synthesized via interfacial self-oxidation–reduction polymerization with $HAuCl_4$ as an aqueous oxidant and the polyaniline as a monomer in the organic toluene layer [35]. These two immiscible layers made contact at an interface; then, the Au^{3+} ions oxidized the aniline monomer to its conducting emeraldine salt polymer formation in the nanotube structure, whereas it was reduced to the nano Au^0 form itself. The AuNP was therefore entrapped on the nanotube surface of the polyaniline, resulting in AuNP-PAni nanocomposites. Finally, the AuNP-PAni nanocomposites were drop-casted on the PAni-coated Au electrode for further analysis.

2.4. Synthesis of N,S-GQD, and Conjugation with Antibody

The synthesis of N,S-GQDs was followed by a hydrothermal reaction system [36]. Antibodies for the target virus were bound with N,S-GQDs using EDC/NHS covalent chemistry [37]. Briefly, 0.1 M EDC was mixed with 5.1 µg of antibody solution and reacted with the carboxyl group contained in the Ab after 30 min of stirring at 7 °C. After that, 1 mL of N,S-GQDs, and 0.1 M NHS were added to activate the amino group on the surface of the GQDs and then stirred for 16 h at 7 °C. The reaction solution was dialyzed by using a 1 kDa dialysis bag to remove the excess EDC and NHS. Finally, the antibody-conjugated N,S-GQD (Ab-N,S-GQDs) solution was stored in 0.1 M PBS (pH 7.4) at 4 °C.

2.5. Fabrication of the Gold Electrode

Deposition of nanocomposites on the gold electrode produces high conductivity to perform electrochemical analysis. Initially, the polyaniline was electrochemically deposited on the gold surface using cyclic voltammetry (CV) in a three-electrode system to prepare the gold electrode. The obtained curve for CV was recorded at a scan rate of 20 mV/s in a potential range of 0–1 V for 10 cycles. After that, 15 µL of the mixed solution of Ab-N,S-GQD with AuNP-PAni was deposited by drop-casting onto the gold surface with polymerized PAni. The sulfur on N,S-GQD formed strong bonds of Au-S with AuNPs via the soft acid–soft base interaction. To minimize the reactivity of the base matrix of Au-PAni, the final sensor electrode was immersed in a solution of 0.2% BSA for blocking before using it for virus detection.

2.6. Preparation of Dengue Virus-Like Particles

The dengue virus-like particles (DENV-LPs) serotypes 1–4 were prepared according to a previously reported protocol [38]. Briefly, the DENV-LPs were expressed in Bm5 cells and purified using affinity chromatography. Transmission electron microscopy revealed that these DENV-LPs formed rough, spherical forms, with a diameter of 30–55 nm. Furthermore, the heparin-binding assay demonstrated that these DENV-LPs contained the envelope protein domain III on their surfaces [39].

2.7. Electrochemical Detection of Virus

The virus solution was diluted in a series of concentrations from 10 fg mL^{-1} to 1 ng mL^{-1} using filtered 0.1 M PBS solution. For detection, 10 µL of virus solution was dropped on the gold electrode containing the Ab-N,S-GQD@AuNP-PAni nanocomposite and incubated for 10 min at room temperature. The virus was bound with the antibody-conjugated surface of the electrode. The unbound virus was washed by dipping the electrode in PBS and then kept in the electrolytic solution. The value for charge transfer resistance (R_{ct}) on the electrode was then measured by the potential EIS with a sinusoidal amplitude of 5 mV within a frequency range from 100 kHz to 0.1 Hz. The virus detection time using the gold electrode was about 15 min.

3. Results

3.1. Preparation of Sensor Electrode and Its Sensing Mechanism

AuNP-PAni nanocomposites were synthesized via interfacial self-oxidation–reduction polymerization with HAuCl$_4$ as an aqueous oxidant and the polyaniline as a monomer in the organic toluene layer. These two layers met at their interfaces; then, the aniline was oxidized to its conducting emeraldine salt polymer formation in the nanotube structure. Au^{3+} was reduced to Au^0, entrapped on the nanotube surface. The synthesis and their TEM images are shown in Figure S1 of the Supplementary material. To synthesize the sensor electrode with the AuNP-PAni nanocomposites, the bare Au electrode was coated with a fine layer of polyaniline via cyclic voltammetry (Figure S2), as presented in the scheme in Figure 1a. The homogeneously distributed N,S-GQDs were prepared by the standard hydrothermal route and conjugated with monoclonal antibodies (Ab) via the EDC/NHS mechanism. The TEM image of the as-synthesized N,S-GQDs are given in Figure S3. Then, the Ab-conjugated N,S-GQD (N,S-GQD-Ab) was dialyzed overnight and drop-casted on the Au|PAni|Au-PAni electrode to synthesize the sensor electrode Au|PAni|Au-PAni|N,S-GQD-Ab. The conjugation between the sulfur atom of N,S-GQD and the AuNP of AuPAni was made by the universal gold–thiol interaction [40,41].

It can be anticipated that the conductivity and the charge storage property of the sensor electrode should possess a high value due to the presence of a conducting surface of AuNP and PAni nanotubes. As shown in the cyclic voltammogram (Figure 1b), the Au|PAni|Au-PAni electrode had a significantly high charge storage capacity with a clear redox peak at +0.45/+0.68 V due to the most electroactive form of the emeraldine salt of polyaniline, compared to the bare Au and Au|PAni electrode [42]. Similarly, its impedance spectrum also exhibits small resistivity in the Nyquist plot in Figure 1c.

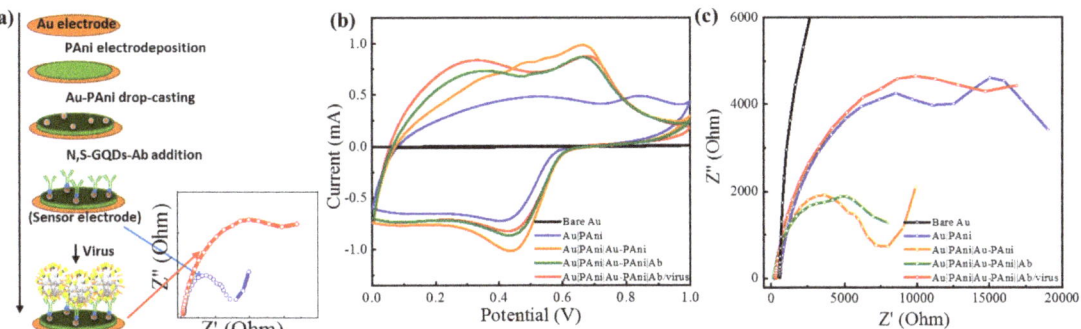

Figure 1. (a) Schematic diagram of the stepwise preparation of the sensor electrode, (b) cyclic voltammetry, and (c) electrochemical impedance results of Au∣PAni∣Au-PAni-Ab electrode before and after virus addition.

3.2. Detection of CHIKV by Au∣PAni∣Au-PAni-N,S-GQD-Ab$_{CHIKV}$ Sensor Electrode

The detection ability was investigated on chikungunya virus with the Au∣PAni∣Au-PAni-N,S-GQD electrode with anti-chikungunya antibody (Ab$_{CHIKV}$). The antibody was conjugated with the N,S-GQDs and dialyzed well before proceeding to Au-PAni-N,S-GQD formation. After that, only sensor electrodes without virus loading were recorded in impedance as the control, which was compared with different virus concentrations loaded on the electrodes. The Nyquist plots of the Au∣PAni∣Au-PAni-N,S-GQD electrode before and after virus loading are shown in Figure 2a, where Z' and Z'' represent the real and imaginary parts, respectively, of the impedance over the frequency range 100 kHz to 100 MHz with an AC amplitude of 5 mV. The plot clearly shows that the increasing pattern in impedance occurred with the increasing virus concentrations from 100 fg·mL^{-1} to 1 ng·mL^{-1}. To observe the individual contributions of impedance value, all Nyquist plots were fitted with several possible equivalent circuit diagrams over frequencies ranging from 100 kHz to 100 MHz. The best-fitted diagram was applied to decipher the individual contribution of the circuit parameters, which is depicted in Table S1. The most crucial factor of these, R_{ct}, represents the transfer of electrons at the electrode to the electrolyte interface. The R_{ct} values found in different concentrations of viruses were compared with their corresponding control values before addition of the virus and are plotted in Figure 2b as the calibration curve. The control value designated in the plot represents the charge transfer between the bare sensor electrode and electrolyte before adding any analyte, which is assigned as 100%. After virus loading, a large number of nonconducting virus molecules covered the conducting surface of N,S-GQDs, and AuNP-PAni, increasing the R_{ct}. The change in R_{ct} was calculated in percentages to obtain the calibration lines. As shown in Figure 2b, the calibration line shows an excellent linear relationship between R_{ct} and the CHIKV concentration. The limit of detection (LOD), determined by $3\sigma/S$ (where σ is the standard deviation of the lowest signal and S is the slope of the calibration line) [25], was found to be 22.1 fg·mL^{-1}.

The main goal of this study was to detect different mosquito-borne viruses. These viruses also have similar surface functionalities that can affect the sensor specificity; thus, it was imperative to investigate the sensor's selectivity, especially with ZIKV and DENV. To analyze the selectivity, the sensor electrode was tested with 10 pg mL^{-1} CHIKV in addition to other samples of ZIKV, DENV-LP-2, Influenza virus A/H1N1, and NoV-LP (all are 10 pg·mL^{-1}). In their Nyquist plots in Figure 2c, it can be seen that the sensor responses to the other viruses are significantly lower due to the nonspecific interaction with the antibody (mentioned in a bracket of Figure 2c), indicating the specificity for the target CHIKV. The R_{ct} values are deciphered from the same circuit diagram, and the percentage change has been plotted in the bar diagram in Figure 2d, where the specificity can be visible.

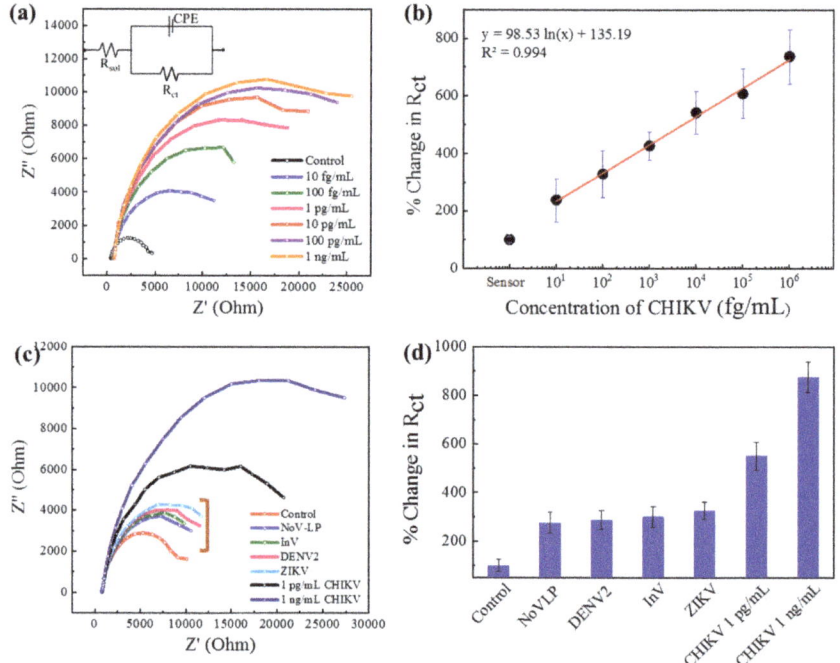

Figure 2. Au|PAni|Au-PAni-N,S-GQD-Ab$_{CHIKV}$ sensor performances in different concentrations of CHKV analytes (10 fg mL^{-1}–1 ng mL^{-1}). (**a**) Nyquist plots (inset: equivalent circuit), (**b**) calibration line for percentage change in R_{ct} vs. CHIKV concentration; selectivity of the proposed sensor with 10 pg mL^{-1} of NoV-LP, DENV-LP-2, Influenzavirus A (H1N1), and ZIKV along with 1 pg mL^{-1} and 1 ng mL^{-1} target CHIKV in (**c**) Nyquist plot and (**d**) bar diagram. Error bars represent the standard deviations of triple measurements.

3.3. Detection of ZIKV by Au|PAni|Au-PAni-N,S-GQD-Ab$_{ZIKV}$ Sensor Electrode

The detection mechanism was based on antibody–antigen interactions; therefore, it can be anticipated that the sensor should be applicable for the detection of other virused by changing the corresponding antibodies on the sensor surface. Therefore, a similar sensor was developed for Zika virus sensing with anti-Zika antibody-conjugated Au-PAni-N,S-GQD nanocomposites. The Nyquist plot and the calibration line show identical results for CHIKV, as presented in Figure 3a. The R_{ct} values for each electrode were deciphered with the same circuit, and the calibration line has been drawn against concentration in Figure 3b. The LODs were also found from the calibration lines, and were as low as 31.1 fg·mL^{-1}, confirming the detection mechanism of EIS-based virus sensing. Similarly, the Au|PAni|Au-PAni-N,S-GQD-Ab$_{ZIKV}$ sensor electrode should possess high selectivity, as shown in the corresponding selectivity values in Figure 3c.

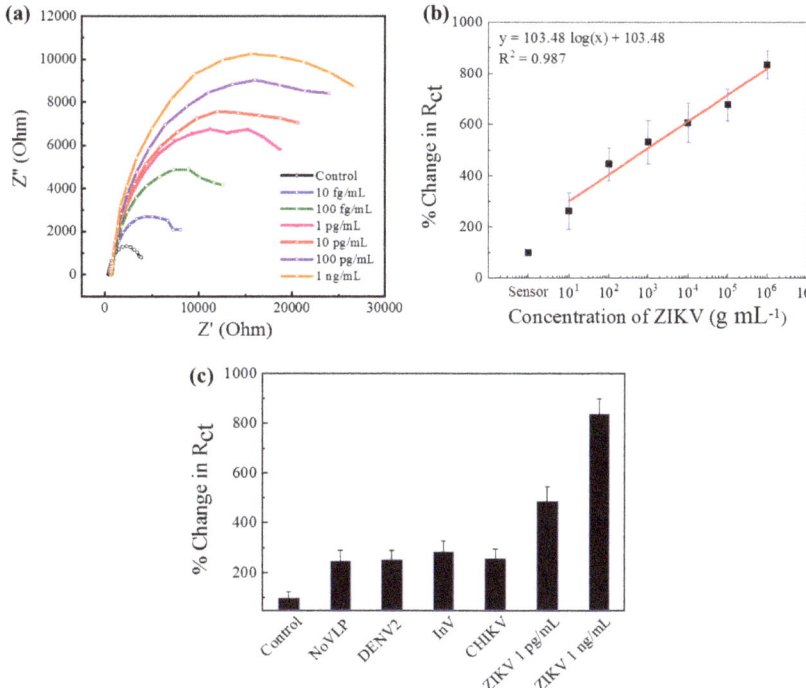

Figure 3. Au | PAni | Au-PAni-N,S-GQD-Ab$_{ZIKV}$ sensor performances in different concentrations of ZIKV analytes (10 fg mL^{-1}–1 ng mL^{-1}). (**a**) Nyquist plots, (**b**) calibration line for percentage change in R_{ct} vs. ZIKV concentration, (**c**) selectivity of the proposed sensor with 10 pg mL^{-1} of NoV-LP, DENV-LP-2, Influenza virus A (H1N1), and ZIKV along with 1 pg mL^{-1}, and 1 ng mL^{-1} target CHIKV. Error bars represent the standard deviations of triple measurements.

3.4. Detection of Dengue Serotypes

Among the mosquito-borne virus diseases, the fatality rate of DENV infection is very high. Treatment becomes critical when secondary infection occurs by a virus serotype different from that of the initial infection [43]. Therefore, serotype identification is equally important as virus detection in the case of DENV. After successfully detecting CHIKV and ZIKV by their corresponding antibody-conjugated Au | PAni | Au-PAni-N,S-GQD sensor electrodes, this proposed method was introduced to detect four serotypes of DENV-LP, separately, for serotyping. The serotypes have many similarities; thus, there was a high possibility for cross-reactivity through the antibody. Therefore, each sensor electrode contained the corresponding monoclonal antibody for serotype sensing. It is clear from Figure 4a–d that the sensor exhibited a significant increment of impedance, increasing with concentration in their Nyquist plots for all the serotypes. Although the increments of the impendence were not of the same magnitude for all serotypes, the trends were very similar for all. The antibody–antigen interactions for different DENV serotypes are not very similar; therefore, it is obvious that virus accumulation on the electrode surface was also slightly different, although treatment with the same concentration of DENV-LP samples resulted in some variation in the impedance value. However, increasing the trend in impedance resulted in a similar concentration response for all serotypes. After measuring their R_{ct} values for the calibration lines in Figure 4e, the trends could be accurately expressed. It can be noted that the slopes of the all-serotypes calibration plots are very similar, which represents their corresponding LOD values of 27.4, 24.5, 41.4, and 13.3 fg mL^{-1} for DENV-LP-1, DENV-LP-2, DENV-LP-3, and DENV-LP-4, respectively.

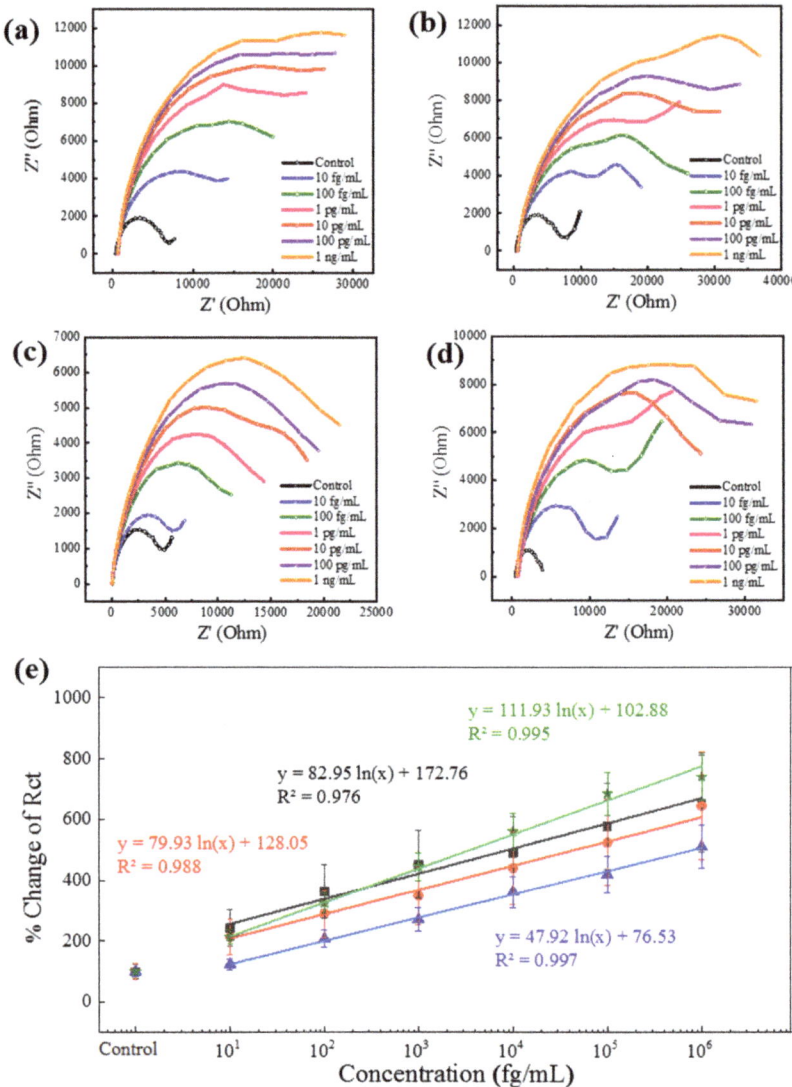

Figure 4. (**a–d**) Nyquist plots of four different DENV-LP serotype antibody-conjugated Au | PAni | Au-PAni-N,S-GQD-Ab$_{DENV}$ electrodes for their corresponding targets in a concentration range from 100 fg mL^{-1} to 1 ng mL^{-1}. (**e**) Calibration lines for all four serotypes in terms of percentage change in R_{ct} vs. DENV-LP concentration. Symbols: squares for DENV-LP-1; circles for DENV-LP-2; stars for DENV-LP-3; and triangles for DENV-LP-4. Error bars represent the standard deviations of triple measurements.

3.5. Performance for Cross-Reactivity of Dengue Serotypes

Although the antibodies used for the sensor preparation were monoclonal, the chances of cross-reactivity were high due to the similarity of viral origin. To check the ability of this sensor to distinguish the identity of different serotypes, four similar electrodes with different antibody-conjugation were applied to verify the cross-reactivity of all four dengue serotypes. The interactions between the four types of sensor electrodes with four serotypes of DENV-LP and four controls are presented in Figure 5. The responses have been converted into the percentage change in R_{ct} for clear understanding. It can be observed from Figure 5a–d that the assigned sensor electrode, which was designed for a specific DENV-LP serotype, was very sensitive for the exact target DENV-LP serotype, showing reduced responses for other serotypes as well as the controls. Control responses for all four sensors were almost insignificant compared with other signals (near 100% for all cases). However, unlike the previous viruses of CHIKV and ZIKV, in these cases, the selectivity showed some positive responses due to the cross-reactive nature of the antibodies. In the Au | PAni | Au-PAni-N,S-GQD-Ab$_{DENV-2}$ sensor electrode, the cross-reactivity showed a 350% enhancement in DENV-LP-1 and four of the target DENV-LP-2, where the actual target of DENV-LP-2 showed a 605% enhancement. The specificity of the antibody for the DENV-LP-2 target was observed to be the most interfering in nature; therefore, the high cross-reactivity of the sensor electrode can be justified [44–46]. These results also proved the successful performances of the sensor electrode, which are highly dependent on the nature of the antibody. However, due to the presence of the antibody, the sensor can possess high specificity but suffers low stability (Figure S4). It is recommended to apply the sensor electrode 1 week within its preparation due to the low stability of the antibody.

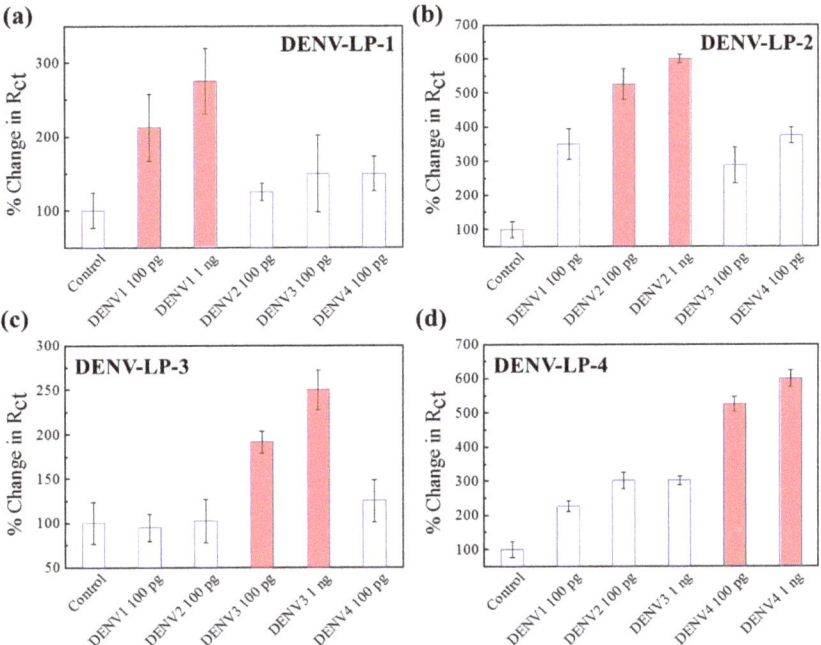

Figure 5. (a–d) Selectivity of four different DENV sensor electrodes with their corresponding target concentrations of 100 pg mL^{-1} and 1 ng mL^{-1} in the presence of other serotypes. Error bars represent the standard deviations of triple measurements.

4. Discussion

This paper proposes an electrochemical biosensor with PAni | Au-PAni-N,S-GQD nanocomposites, combining different antibodies to detect their corresponding viruses. The AuNP-PAni nanocomposites that conjugated N,S-GQDs-Ab enhanced the electron transfer process, which improved the electrochemical response and provided a small resistive value in the electrochemical impedance spectroscopy. The Au nanoparticles (AuNPs) with well-defined and controlled shapes incorporated in the nanochain of a conducting polymer, polyaniline, have attracted increasing attention as a promising material for biosensing matrixes. On the other hand, N,S-GQDs with structural defects in N and S can present useful functionality for nanocomposite formation. In N,S-GQDs, the nitrogen atom drastically enhances the electrochemical properties of GQD. In contrast, sulfur can increase the coordinate binding with the AuNP, situated in the Au-PAni nanocomposites. The N,S-GQDs can conjugate with the antibody embedded with Au-PAni. Therefore, the Ab-N,S-GQDs@AuNP-PAni nanocomposites show excellent electroactivity in solution, and can be applied for impedimetric virus detection. After the addition of the virus, the sensor electrode can bind with the viruses due to the conjugated antibody on the electrode surface, where the R_{ct} value between the sensor electrode and the electrolyte solution exhibits a significant change, resulting in a large increase in the EIS result. Due to the usage of antibody–antigen interactions on the nanocomposites, the sensor shows excellent selectivity and minimal cross-reactivity in the presence of other viruses. In the case of DENV serotyping, where the possibility of cross-reactivity is very high, this sensor can identify the serotype in the concentration of pg mL^{-1}, which to the best of our knowledge, is the first attempt using an electrochemical process. In recent studies, the identification of serotypes has not been investigated in DENV detection. As listed in Table 1, most reports mainly focus on DENV detection through the NS1 protein or secondary antibody concentrations [47,48]. However, it is always better to detect the direct presence of virus concentrations rather than proteins or IgG or IgM, because their concentrations in the initial stage of infection are significantly low compared to direct virus loading. In serotyping, few reports have been published targeting the oligomers of different serotypes where the extraction of the RNA is a time-consuming process. Observing the overall performance, although the cross-reactivity was not negligible, we could still confirm qualitative information about the DENV serotype, which is highly necessary for DENV diagnosis.

Table 1. Comparison of the detectability of the proposed sensor with recently reported DENV sensors.

Detection Method	Analytes	LOD	Detection Range	References
Electrochemical	DENV NS 1 protein	1.49 µg mL^{-1}	0–1.4 µg mL^{-1}	[49]
SPR—optical	DENV type 2 E proteins	0.08 pM	0.08–0.5 pM	[50]
Optical DNA biosensor	DENV serotype 2	10^{-21} M	1.0×10^{-15}–1.0×10^{-11} M	[51]
SRP—biosensor	DENV serotype 2 and 3	2×10^4 particles mL^{-1}	–	[52]
Colorimetric	Different DENV serotype	–	–	[48]
Fluorometric	DENV all serotypes	9.4 fM	10^{-14} to 10^{-6} M	[25]
SERS-based lateral flow biosensor	DENV NS 1 protein	15 ng mL^{-1}	15–500 ng mL^{-1}	[53]
Electrochemical	DENV NS 1 protein	0.3 ng mL^{-1}	1–200 ng mL^{-1}	[54]
Electrochemical	DENV-LP 1 serotype	27.4 fg mL^{-1}	100 fg^{-1} ng mL^{-1}	This work
Electrochemical	DENV-LP 2 serotype	24.5 fg mL^{-1}	100 fg^{-1} ng mL^{-1}	This work
Electrochemical	DENV-LP 3 serotype	41.4 fg mL^{-1}	100 fg^{-1} ng mL^{-1}	This work
Electrochemical	DENV-LP 4 serotype	13.3 fg mL^{-1}	100 fg^{-1} ng mL^{-1}	This work

5. Conclusions

During the past few decades, several investigations have been carried out to establish a sensitive detection technique of viruses. Although few biosensors have improved virus detection in selectivity, sensitivity, and response time, practical usages are limited, especially in cases where the analytes derive from similar origins. This report has successfully developed an electrochemical biosensor with PAni | Au-PAni-N,S-GQD nanocomposites, combining different antibodies to detect their corresponding viruses. We have targeted the mosquito-borne viruses DENV-1, DENV-2, DENV-3, DENV-4, CHIKV, and ZIKV in their detection, conjugating their corresponding antibodies on the nanocomposites. In all cases, we achieved high sensitivity, with LOD values of 22.1, 31.1, 27.4, 24.5, 41.4, and 13.3 fg mL^{-1} for CHIKV, ZIKV, DENV-LP-1, DENV-LP-2, DENV-LP-3, and DENV-LP-4, respectively. We hope that the proper development of this method for applications in disposable and multiplex systems can result in a single sensor to detect several closely related viruses in the near future.

Supplementary Materials: The following are available online at https://www.mdpi.com/article/10.3390/bios11100376/s1, Table S1: (a) Circuit diagram for electrochemical fitting, (b) electrochemical parameters of the sensor electrode obtained from the impedimetric circuit diagram, Figure S1: (a) Schematic illustration of the Au-PAni-N,S-GQD-Ab sensor electrode, (b) TEM image of the Au-PAni nanocomposite, Figure S2: Electropolymerization of polyaniline, Figure S3: TEM image of N,S-GQDs.

Author Contributions: Conceptualization, F.N., A.D.C. and E.Y.P.; methodology, F.N., K.T. and D.I.S.U.; validation, F.N., A.D.C. and E.Y.P.; formal analysis, F.N., K.T. and D.I.S.U.; investigation, F.N., A.D.C. and E.Y.P.; writing—original draft preparation, F.N. and A.D.C.; writing—review and editing, A.D.C. and E.Y.P.; supervision, E.Y.P.; funding acquisition, E.Y.P. All authors have read and agreed to the published version of the manuscript.

Funding: This work was partially supported by the Fund for the Promotion of Joint International Research, Fostering Joint International Research B (Grant No. 20KK0115), and the Japan Agency for Medical Research and Development (20hm0102080h0001).

Institutional Review Board Statement: Not applicable.

Informed Consent Statement: Not applicable.

Data Availability Statement: The supporting data for this study are available from the corresponding author upon reasonable request.

Acknowledgments: The authors thank K. S. Boo (Insect Pathology Laboratory, School of Agricultural Biotechnology, Seoul National University, Seoul, South Korea) and Tian-Cheng Li (National Institute of Infectious Diseases) for providing the Bm5 cells and norovirus-like particles, respectively.

Conflicts of Interest: The authors declare no conflict of interest.

References

1. WHO. Vector-Borne Diseases. Available online: https://www.who.int/news-room/fact-sheets/detail/vector-borne-diseases#:~{}:text=Vector%2Dborne%20diseases%20account%20for,infection%20transmitted%20by%20Anopheline%20mosquitoes (accessed on 24 July 2021).
2. Hayes, E.B. Zika virus outside Africa. *Emerg. Infect. Dis.* **2009**, *15*, 1347. [CrossRef]
3. Luo, L.; Jiang, L.-Y.; Xiao, X.-C.; Di, B.; Jing, Q.-L.; Wang, S.-Y.; Tang, J.-L.; Wang, M.; Tang, X.-P.; Yang, Z.-C. The dengue preface to endemic in mainland China: The historical largest outbreak by Aedes albopictus in Guangzhou, 2014. *Infect. Dis. Poverty* **2017**, *6*, 148. [CrossRef]
4. Qiaoli, Z.; Jianfeng, H.; De, W.; Zijun, W.; Xinguang, Z.; Haojie, Z.; Fan, D.; Zhiquan, L.; Shiwen, W.; Zhenyu, H. Maiden outbreak of chikungunya in Dongguan city, Guangdong province, China: Epidemiological characteristics. *PLoS ONE* **2012**, *7*, e42830. [CrossRef] [PubMed]
5. Beltrán-Silva, S.; Chacón-Hernández, S.; Moreno-Palacios, E.; Pereyra-Molina, J. Clinical and differential diagnosis: Dengue, chikungunya and Zika. *Rev. Med. del Hosp. Gen. Mex.* **2018**, *81*, 146–153. [CrossRef]
6. Mayer, S.V.; Tesh, R.B.; Vasilakis, N. The emergence of arthropod-borne viral diseases: A global prospective on dengue, chikungunya and zika fevers. *Acta Trop.* **2017**, *166*, 155–163. [CrossRef] [PubMed]
7. Wasserman, S.; Tambyah, P.A.; Lim, P.L. Yellow fever cases in Asia: Primed for an epidemic. *Int. J. Infect. Dis.* **2016**, *48*, 98–103. [CrossRef]
8. Pongsiri, P.; Praianantathavorn, K.; Theamboonlers, A.; Payungporn, S.; Poovorawan, Y. Multiplex real–time RT–PCR for detecting chikungunya virus and dengue virus. *Asian Pac. J. Trop Dis.* **2012**, *5*, 342–346. [CrossRef]
9. Wu, W.; Wang, J.; Yu, N.; Yan, J.; Zhuo, Z.; Chen, M.; Su, X.; Fang, M.; He, S.; Zhang, S. Development of multiplex real-time reverse–transcriptase polymerase chain reaction assay for simultaneous detection of Zika, dengue, yellow fever, and chikungunya viruses in a single tube. *J. Med. Virol.* **2018**, *90*, 1681–1686. [CrossRef]
10. WHO. Dengue and Severe Dengue. 2021. Available online: https://www.who.int/news-room/fact-sheets/detail/dengue-and-severe-dengue (accessed on 25 July 2021).
11. Li, J.; Xiong, Y.; Wu, W.; Liu, X.; Qu, J.; Zhao, X.; Zhang, S.; Li, J.; Li, W.; Liao, Y. Zika virus in a traveler returning to China from Caracas, Venezuela, February 2016. *Emerg. Infect. Dis.* **2016**, *22*, 1133. [CrossRef]

12. Wang, B.; Liang, Y.; Lu, Y.; Zhang, L.; Li, Y.; Song, Y.; Qin, C.; Luo, Z.; Xia, Z.; Qin, W. The importation of the phylogenetic-transition state of Zika virus to China in 2014. *J. Infect.* **2018**, *76*, 106–109. [CrossRef]
13. Payungporn, S.; Chutinimitkul, S.; Chaisingh, A.; Damrongwantanapokin, S.; Buranathai, C.; Amonsin, A.; Theamboonlers, A.; Poovorawan, Y. Single step multiplex real-time RT-PCR for H5N1 influenza A virus detection. *J. Virol. Methods* **2006**, *131*, 143–147. [CrossRef] [PubMed]
14. Singanayagam, A.; Patel, M.; Charlett, A.; Bernal, J.L.; Saliba, V.; Ellis, J.; Ladhani, S.; Zambon, M.; Gopal, R. Duration of infectiousness and correlation with RT-PCR cycle threshold values in cases of COVID-19, England, January to May 2020. *Eurosurveillance* **2020**, *25*, 2001483. [CrossRef] [PubMed]
15. Fischer, K.; Diederich, S.; Smith, G.; Reiche, S.; Pinho dos Reis, V.; Stroh, E.; Groschup, M.H.; Weingartl, H.M.; Balkema-Buschmann, A. Indirect ELISA based on Hendra and Nipah virus proteins for the detection of henipavirus specific antibodies in pigs. *PLoS ONE* **2018**, *13*, e0194385. [CrossRef]
16. Shukla, S.; Hong, S.-Y.; Chung, S.H.; Kim, M. Rapid detection strategies for the global threat of Zika virus: Current state, new hypotheses, and limitations. *Front. Microbiol.* **2016**, *7*, 1685. [CrossRef] [PubMed]
17. Yamada, K.-I.; Takasaki, T.; Nawa, M.; Kurane, I. Virus isolation as one of the diagnostic methods for dengue virus infection. *J. Clin. Virol.* **2002**, *24*, 203–209. [CrossRef]
18. Domingo, C.; Niedrig, M.; Teichmann, A.; Kaiser, M.; Rumer, L.; Jarman, R.G.; Donoso-Mantke, O. 2 nd international external quality control assessment for the molecular diagnosis of dengue infections. *PLoS Negl. Trop. Dis.* **2010**, *4*, e833. [CrossRef]
19. Chowdhury, A.D.; Sharmin, S.; Nasrin, F.; Yamazaki, M.; Abe, F.; Suzuki, T.; Park, E.Y. Use of Target-Specific Liposome and Magnetic Nanoparticle Conjugation for the Amplified Detection of Norovirus. *ACS Appl. Bio Mater.* **2020**, *3*, 3560–3568. [CrossRef]
20. Chowdhury, A.D.; Takemura, K.; Li, T.-C.; Suzuki, T.; Park, E.Y. Electrical pulse-induced electrochemical biosensor for hepatitis E virus detection. *Nature Commun.* **2019**, *10*, 3737. [CrossRef]
21. Boonham, N.; Kreuze, J.; Winter, S.; van der Vlugt, R.; Bergervoet, J.; Tomlinson, J.; Mumford, R. Methods in virus diagnostics: From ELISA to next generation sequencing. *Virus Res.* **2014**, *186*, 20–31. [CrossRef] [PubMed]
22. Xia, Y.; Chen, Y.; Tang, Y.; Cheng, G.; Yu, X.; He, H.; Cao, G.; Lu, H.; Liu, Z.; Zheng, S.-Y. Smartphone-based point-of-care microfluidic platform fabricated with a ZnO nanorod template for colorimetric virus detection. *ACS Sens.* **2019**, *4*, 3298–3307. [CrossRef] [PubMed]
23. Giry, C.; Roquebert, B.; Li-Pat-Yuen, G.; Gasque, P.; Jaffar-Bandjee, M.-C. Simultaneous detection of chikungunya virus, dengue virus and human pathogenic Leptospira genomes using a multiplex TaqMan®assay. *BMC Microbiol.* **2017**, *17*, 105. [CrossRef]
24. Chowdhury, A.D.; Takemura, K.; Khorish, I.M.; Nasrin, F.; Tun, M.M.N.; Morita, K.; Park, E.Y. The detection and identification of dengue virus serotypes with quantum dot and AuNP regulated localized surface plasmon resonance. *Nanoscale Adv.* **2020**, *2*, 699–709. [CrossRef]
25. Dutta Chowdhury, A.; Ganganboina, A.B.; Nasrin, F.; Takemura, K.; Doong, R.-A.; Utomo, D.I.S.; Lee, J.; Khoris, I.M.; Park, E.Y. Femtomolar detection of dengue virus DNA with serotype identification ability. *Anal. Chem.* **2018**, *90*, 12464–12474. [CrossRef]
26. Ganganboina, A.B.; Chowdhury, A.D.; Khoris, I.M.; Doong, R.-A.; Li, T.-C.; Hara, T.; Abe, F.; Suzuki, T.; Park, E.Y. Hollow magnetic-fluorescent nanoparticles for dual-modality virus detection. *Biosen. Bioelectron.* **2020**, *170*, 112680. [CrossRef]
27. Nasrin, F.; Chowdhury, A.D.; Ganganboina, A.B.; Achadu, O.J.; Hossain, F.; Yamazaki, M.; Park, E.Y. Fluorescent and electrochemical dual-mode detection of Chikungunya virus E1 protein using fluorophore-embedded and redox probe-encapsulated liposomes. *Microchim. Acta* **2020**, *187*, 674. [CrossRef]
28. Li, Q.; Wu, J.-T.; Liu, Y.; Qi, X.-M.; Jin, H.-G.; Yang, C.; Liu, J.; Li, G.-L.; He, Q.-G. Recent advances in black phosphorus-based electrochemical sensors: A review. *Anal. Chim. Acta* **2021**, *1170*, 338480. [CrossRef]
29. Kirchhain, A.; Bonini, A.; Vivaldi, F.; Di Francesco, F. Latest developments in non-faradic impedimetric biosensors: Towards clinical applications. *TrAC Trends Anal. Chem.* **2020**, *133*, 116073. [CrossRef]
30. Li, Q.; Xia, Y.; Wan, X.; Yang, S.; Cai, Z.; Ye, Y.; Li, G. Morphology-dependent MnO2/nitrogen-doped graphene nanocomposites for simultaneous detection of trace dopamine and uric acid. *Mater. Sci. Eng. C* **2020**, *109*, 110615. [CrossRef]
31. Ganganboina, A.B.; Doong, R.-a. Functionalized N-doped graphene quantum dots for electrochemical determination of cholesterol through host-guest inclusion. *Microchim. Acta* **2018**, *185*, 526. [CrossRef]
32. Nithyadharseni, P.; Reddy, M.; Nalini, B.; Kalpana, M.; Chowdari, B.V. Sn-based intermetallic alloy anode materials for the application of lithium ion batteries. *Electrochim. Acta* **2015**, *161*, 261–268. [CrossRef]
33. Reddy, M.; Wei Wen, B.L.; Loh, K.P.; Chowdari, B. Energy storage studies on InVO4 as high performance anode material for Li-ion batteries. *ACS Appl. Mater. Interfaces* **2013**, *5*, 7777–7785. [CrossRef]
34. Ganganboina, A.B.; Dutta Chowdhury, A.; Doong, R.-a. New avenue for appendage of graphene quantum dots on halloysite nanotubes as anode materials for high performance supercapacitors. *ACS Sustain. Chem. Eng.* **2017**, *5*, 4930–4940. [CrossRef]
35. Chowdhury, A.D.; Gangopadhyay, R.; De, A. Highly sensitive electrochemical biosensor for glucose, DNA and protein using gold-polyaniline nanocomposites as a common matrix. *Sens. Actuators B* **2014**, *190*, 348–356. [CrossRef]
36. Ganganboina, A.B.; Doong, R.-A. Graphene quantum dots decorated gold-polyaniline nanowire for impedimetric detection of carcinoembryonic antigen. *Sci. Rep.* **2019**, *9*, 7214. [CrossRef]
37. Raghav, R.; Srivastava, S. Immobilization strategy for enhancing sensitivity of immunosensors: L-Asparagine–AuNPs as a promising alternative of EDC–NHS activated citrate–AuNPs for antibody immobilization. *Biosen. Bioelectron.* **2016**, *78*, 396–403. [CrossRef] [PubMed]

38. Utomo, D.I.S.; Pambudi, S.; Sjatha, F.; Kato, T.; Park, E.Y. Production of dengue virus-like particles serotype-3 in silkworm larvae and their ability to elicit a humoral immune response in mice. *AMB Express* **2020**, *10*, 147. [CrossRef]
39. Utomo, D.I.S.; Hirono, I.; Kato, T.; Park, E.Y. Formation of virus-like particles of the dengue virus serotype 2 expressed in silkworm larvae. *Mol. Biotechnol.* **2019**, *61*, 852–859. [CrossRef]
40. Mahmoud, A.M.; El-Wekil, M.M.; Mahnashi, M.H.; Ali, M.F.; Alkahtani, S.A. Modification of N, S co-doped graphene quantum dots with p-aminothiophenol-functionalized gold nanoparticles for molecular imprint-based voltammetric determination of the antiviral drug sofosbuvir. *Microchim. Acta* **2019**, *186*, 617. [CrossRef]
41. Yao, J.; Li, Y.; Xie, M.; Yang, Q.; Liu, T. The electrochemical behaviors and kinetics of AuNPs/N, S-GQDs composite electrode: A novel label-free amplified BPA aptasensor with extreme sensitivity and selectivity. *J. Mol. Liq.* **2020**, *320*, 114384. [CrossRef]
42. Song, E.; Choi, J.-W. Conducting polyaniline nanowire and its applications in chemiresistive sensing. *Nanomaterials* **2013**, *3*, 498–523. [CrossRef]
43. Soo, K.-M.; Khalid, B.; Ching, S.-M.; Chee, H.-Y. Meta-analysis of dengue severity during infection by different dengue virus serotypes in primary and secondary infections. *PLoS ONE* **2016**, *11*, e0154760.
44. Midgley, C.M.; Flanagan, A.; Tran, H.B.; Dejnirattisai, W.; Chawansuntati, K.; Jumnainsong, A.; Wongwiwat, W.; Duangchinda, T.; Mongkolsapaya, J.; Grimes, J.M. Structural analysis of a dengue cross-reactive antibody complexed with envelope domain III reveals the molecular basis of cross-reactivity. *J. Immunol.* **2012**, *188*, 4971–4979. [CrossRef]
45. Lok, S.-M.; Kostyuchenko, V.; Nybakken, G.E.; Holdaway, H.A.; Battisti, A.J.; Sukupolvi-Petty, S.; Sedlak, D.; Fremont, D.H.; Chipman, P.R.; Roehrig, J.T. Binding of a neutralizing antibody to dengue virus alters the arrangement of surface glycoproteins. *Nat. Struct. Mol. Biol.* **2008**, *15*, 312–317. [CrossRef]
46. Luna, D.M.; Avelino, K.Y.; Cordeiro, M.T.; Andrade, C.A.; Oliveira, M.D. Electrochemical immunosensor for dengue virus serotypes based on 4-mercaptobenzoic acid modified gold nanoparticles on self-assembled cysteine monolayers. *Sens. Actuators B* **2015**, *220*, 565–572. [CrossRef]
47. Vinayagam, S.; Rajaiah, P.; Mukherjee, A.; Natarajan, C. DNA-triangular silver nanoparticles nanoprobe for the detection of dengue virus distinguishing serotype. *Spectrochim. Acta A Mol. Biomol. Spectrosc.* **2018**, *202*, 346–351. [CrossRef]
48. Dutta, R.; Thangapandi, K.; Mondal, S.; Nanda, A.; Bose, S.; Sanyal, S.; Jana, S.K.; Ghorai, S. Polyaniline based electrochemical sensor for the detection of dengue virus infection. *Avicenna J. Med Biotechnol.* **2020**, *12*, 77.
49. Kim, J.H.; Cho, C.H.; Ryu, M.Y.; Kim, J.-G.; Lee, S.-J.; Park, T.J.; Park, J.P. Development of peptide biosensor for the detection of dengue fever biomarker, nonstructural 1. *PLoS ONE* **2019**, *14*, e0222144. [CrossRef]
50. Omar, N.A.S.; Fen, Y.W.; Abdullah, J.; Kamil, Y.M.; Daniyal, W.M.E.M.M.; Sadrolhosseini, A.R.; Mahdi, M.A. Sensitive detection of dengue virus type 2 E-proteins signals using self-assembled monolayers/reduced graphene oxide-PAMAM dendrimer thin film-SPR optical sensor. *Sci. Rep.* **2020**, *10*, 2374. [CrossRef]
51. Mazlan, N.-F.; Tan, L.L.; Karim, N.H.A.; Heng, L.Y.; Reza, M.I.H. Optical biosensing using newly synthesized metal salphen complexes: A potential DNA diagnostic tool. *Sens. Actuators B* **2017**, *242*, 176–188. [CrossRef]
52. Loureiro, F.C.; Neff, H.; Melcher, E.U.; Roque, R.A.; de Figueiredo, R.M.; Thirstrup, C.; Borre, M.B.; Lima, A.M. Simplified immunoassay for rapid Dengue serotype diagnosis, revealing insensitivity to non-specific binding interference. *Sens. Biosens. Res.* **2017**, *13*, 96–103. [CrossRef]
53. Darwish, N.T.; Sekaran, S.D.; Alias, Y.; Khor, S.M. Immunofluorescence–based biosensor for the determination of dengue virus NS1 in clinical samples. *J. Pharm. Biomed. Anal.* **2018**, *149*, 591–602. [CrossRef]
54. Nawaz, M.H.; Hayat, A.; Catanante, G.; Latif, U.; Marty, J.L. Development of a portable and disposable NS1 based electrochemical immunosensor for early diagnosis of dengue virus. *Anal. Chim. Acta* **2018**, *1026*, 1–7. [CrossRef]

 *biosensors*

Article

A Single-Substrate Biosensor with Spin-Coated Liquid Crystal Film for Simple, Sensitive and Label-Free Protein Detection

Po-Chang Wu [1], Chao-Ping Pai [1], Mon-Juan Lee [2,3,*] and Wei Lee [1,*]

[1] Institute of Imaging and Biomedical Photonics, College of Photonics, National Yang Ming Chiao Tung University, Guiren Dist., Tainan 711010, Taiwan; jackywu@nycu.edu.tw (P.-C.W.); pdxdydz@gmail.com (C.-P.P.)
[2] Department of Bioscience Technology, Chang Jung Christian University, Guiren Dist., Tainan 711301, Taiwan
[3] Department of Medical Science Industries, Chang Jung Christian University, Guiren Dist., Tainan 711301, Taiwan
* Correspondence: mjlee@mail.cjcu.edu.tw (M.-J.L.); wei.lee@nycu.edu.tw (W.L.)

Abstract: A liquid crystal (LC)-based single-substrate biosensor was developed by spin-coating an LC thin film on a dimethyloctadecyl[3-(trimethoxysilyl)propyl]ammonium chloride (DMOAP)-decorated glass slide. Compared with the conventional sandwiched cell configuration, the simplified procedure for the preparation of an LC film allows the film thickness to be precisely controlled by adjusting the spin rate, thus eliminating personal errors involved in LC cell assembly. The limit of detection (LOD) for bovine serum albumin (BSA) was lowered from 10^{-5} g/mL with a 4.2-μm-thick sandwiched cell of the commercial LC E7 to 10^{-7} g/mL with a 4.2-μm-thick spin-coated E7 film and further to 10^{-8} g/mL by reducing the E7 film thickness to 3.4 μm. Moreover, by exploiting the LC film of the highly birefringent nematic LC HDN in the immunodetection of the cancer biomarker CA125, an LOD comparable to that determined with a sandwiched HDN cell was achieved at 10^{-8} g/mL CA125 using a capture antibody concentration an order of magnitude lower than that in the LC cell. Our results suggest that employing spin-coated LC film instead of conventional sandwiched LC cell provides a more reliable, reproducible, and cost-effective single-substrate platform, allowing simple fabrication of an LC-based biosensor for sensitive and label-free protein detection and immunoassay.

Keywords: liquid crystal; spin-coating; single-substrate; label-free biosensor; bovine serum albumin; cancer biomarker CA125

Citation: Wu, P.-C.; Pai, C.-P.; Lee, M.-J.; Lee, W. A Single-Substrate Biosensor with Spin-Coated Liquid Crystal Film for Simple, Sensitive and Label-Free Protein Detection. *Biosensors* **2021**, *11*, 374. https://doi.org/10.3390/bios11100374

Received: 1 August 2021
Accepted: 3 October 2021
Published: 6 October 2021

Publisher's Note: MDPI stays neutral with regard to jurisdictional claims in published maps and institutional affiliations.

Copyright: © 2021 by the authors. Licensee MDPI, Basel, Switzerland. This article is an open access article distributed under the terms and conditions of the Creative Commons Attribution (CC BY) license (https://creativecommons.org/licenses/by/4.0/).

1. Introduction

Liquid crystals (LCs) have been extensively exploited as a sensing element for biological detections since Abbott et al. first demonstrated in 1998 the use of the well-known single compound LC 5CB to transduce and amplify the optical signal produced by ligand-receptor binding at LC-solid interfaces for the detection of avidin [1]. By virtue of the unique properties of LCs, including optical anisotropy, fast stimuli-responsive molecular orientation, long-range anchoring transition and short-range intermolecular interaction, the LC-based biosensing mechanism is based on the presence of proteins or biological binding events at the interface (e.g., LC-aqueous or LC-solid) that is capable of reorienting LC molecules, typically from a uniform homeotropic or planar state to a disrupted state [2]. Such a response in LC orientation can then be transduced into a visible optical signal to the naked eye by observing the LC texture under crossed polarizers, allowing label-free detection to be achieved with the advantages of high sensitivity, rapid response, low cost, and simple operation. A wide variety of LC-based biosensing technologies were demonstrated at the LC-solid or LC-aqueous interface for different types of biological analytes, with strategies to improve sensing performance summarized in several review papers [3–9]. While conventional optical texture observation mainly permits qualitative analysis of the optical signal, attention has been paid to developing quantitative biosensing techniques by exploiting the dielectric and electro-optical characteristics of nematic LCs [10–12], Bragg reflection of chiral LCs [13–15], and selective absorption features of a dye-LC as well as a dye-doped LC [16,17].

Because of its fluidity, LC is typically confined in a well-defined compartment for biosensing applications. Conventionally, an LC-aqueous interface is established by filling LCs in a transmission electron microscopy (TEM) grid mesh, with homeotropic and planar anchoring at the LC-air and LC-water interface, respectively. In the solid-LC-aqueous configuration, the TEM grid was placed on a glass substrate coated with silane coupling agents (e.g., dimethyloctadecyl[3-(trimethoxysilyl)propyl]ammonium chloride, DMOAP or octadecyltrichlorosilane, OTS) and impregnated with LC, followed by immersion in an aqueous solution containing the analyte [18,19], while the air-LC-aqueous configuration was constituted by layering the LC-filled TEM grid on top of and in contact with the aqueous phase, keeping the other side of the LC film exposed to the air [20,21]. On the other hand, most biosensing platforms relying on detection at the LC-solid interface are developed with a sandwiched LC cell configuration, in which the LC is enclosed between a pair of glass substrates with the inner surfaces coated with DMOAP or OTS to support homeotropic anchoring of LC molecules. Therefore, the presence of biomolecules on one of the substrate surfaces can be detected with high signal-to-background contrast by the dark-to-bright transition of the optical LC texture when the homeotropic alignment is disrupted [22]. However, fabrication of an LC-based biosensor with a sandwiched LC cell or LC-filled TEM grid is time-consuming and requires trained personnel, and procedures such as the construction of an LC cell (e.g., spacer dispersion, adhesive sealing, and substrates assembly) may introduce personal errors in the uniformity and thickness of the LC layer, which would reduce sample-to-sample reproducibility as well as accuracy and reliability of an LC-based bioassay. As such, LC-based sensing platforms eliminating cell assembly have been proposed, such as injecting LCs in microfluidic channels [23,24] or rectangular capillaries [25] for the detection of antibody-antigen immunobinding on a solid surface or dispensing LC-in-water droplet patterns on an OTS-treated substrate for detection in an aqueous solution [26,27].

Along the line of simplifying the procedure of fabrication, in this study we proposed to spin-coat LCs in the form of a thin film directly on a DMOAP glass substrate and exploited this single-substrate configuration as the sensing platform to report the presence of protein or the occurrence of specific antibody-antigen interactions on the solid surface. Spin-coating is a mature manufacturing process in the microelectronics and semiconductor industries that utilizes the centrifugal force to simply and rapidly deposit a thin film on a flat surface with thickness ranging from micro- to nanometers, which is controllable by adjusting the spin rate. So far, the spin-coating technique has been widely applied to support film formation of photo-resistant, polymeric, and semiconducting materials for optical, electronic, solar cell, semiconducting, display, and sensing applications. In contrast to traditional sandwiched LC cells, the spin-coating procedure can be easily familiarized by an inexperienced user to obtain an LC film. Spin-coated LC film facilitates fundamental investigations on LC phase transition and morphological transformation [28,29], as well as light-driven helical rotation and pitch tuning of cholesteric LC gratings [30]. In our previous studies, the single-substrate biosensing technology was reported with spin-coated films of an LC-photopolymer composite and cholesteric LC [15,31]. Herein, we extend the application of single-substrate detection to nematic LC films in protein assay with bovine serum albumin (BSA) as the protein standard and immunoassay of the cancer biomarker CA125. Results were compared with those using a sandwiched LC cell and approaches to signal amplification concerning LC film thickness were discussed.

2. Materials and Methods

2.1. Materials

The NEG AT35 optical glass slides with the dimensions of 22 mm in length, 18 mm in width and 1.1 mm in thickness were received from Ruilong Glass, Taiwan. The aligning agent DMOAP was purchased from Sigma-Aldrich (St. Louis, MO, USA). The protein standard BSA (Sigma-Aldrich, St. Louis, MO, USA) with a molecular weight of 66.43 kDa was adopted in protein assays, while recombinant human CA125 (MUC16) protein received from R&D Systems (Minneapolis, MN, USA) and anti-CA125 antibody provided by Santa Cruz Biotechnology (Dallas, TX, USA) were used in immunoassays. The two eutectic nematic LCs, E7 and HDN, were obtained from Daily Polymer Co., Taiwan, and Jiangsu Hecheng Display Technology Co., China, respectively. Their clearing temperatures (T_c) are 58 °C and 97 °C, while birefringence (Δn) measured at the wavelength of 589 nm and temperature of 20 °C is 0.225 and 0.333 for E7 and HDN, respectively. Deionized (DI) water, purified by an RDI reverse osmosis/deionizer system, was used to prepare all aqueous solutions.

2.2. Formation of DMOAP Monolayer

DMOAP-coated substrates were prepared to bear immobilized biomolecules (e.g., BSA and anti-CA125 antibodies) and to support homeotropic LC orientation at the LC-glass interface. Following the procedure for cleaning purposes established in our previous works, optical glass slides prior to use in experiments were washed under ultrasonication with a sequence of an aqueous solution of detergent, DI water and 99% ethanol. After performing each of the above-mentioned steps at room temperature for 15 min, cleaned glass slides were dried with nitrogen, baked in an oven at 74 °C for 15 min and then cooled down naturally to room temperature. Using the dip-coating method, cleaned glass slides were immersed and ultrasonicated in an aqueous solution containing 1% (v/v) DMOAP for 15 min. After washing in DI water for 15 min and drying under a stream of nitrogen, these slides were baked in an oven at 85 °C for 15 min to form a stable and uniform DMOAP monolayer on the glass surface.

2.3. Immobilization of BSA

Aqueous solutions with designated BSA concentrations were prepared by serial dilution of a BSA stock solution with DI water. A DMOAP-coated glass slide was dispensed with four droplets (5 μL/droplet) of BSA solution at a given concentration using a micropipette to form a 2 × 2 protein array. After incubation at 30 °C for 1 h to allow immobilization of BSA, the slide was rinsed twice with DI water to remove unbound BSA and then incubated in an oven at 30 °C for 30 min to allow evaporation of DI water without affecting BSA activity.

2.4. Specific Binding of Anti-CA125 Antibody to the CA125 Antigen

Anti-CA125 antibody was diluted to the desired concentration with DI water, while the lyophilized powder of CA125 was reconstituted in phosphate buffered saline followed by serial dilution in DI water. To perform a CA125 immunoassay, anti-CA125 antibody was immobilized at 5 μL/droplet on a DMOAP-coated substrate in a 2 × 2 array format. After drying for 1 h and rinsing twice thoroughly with DI water, the substrate was dispensed with 15 μL CA125 antigen and covered with a cleaned cover glass to allow specific binding of CA125 to the immobilized anti-CA125 antibody for 30 min. After removing the cover glass, the substrate was rinsed twice with DI water to eliminate unbound CA125 and then dried in an oven at 30 °C.

2.5. LC Cell Assembly and Spin-Coating of LC Films

An LC cell was fabricated by assembling a BSA-immobilized substrate and another BSA-free DMOAP-coated slide, with the DMOAP coating and immobilized BSA facing inward, to form a sandwiched cell of 4.2 ± 0.5 μm (determined by the size of ball spacers) in cell gap, which was then filled with LC through capillary force. LC films were formed by dispensing 5 μL of E7 or HDN on areas on the glass substrate immobilized with the analyte, followed by spin-coating with a SP-D$_3$-P spin coater (APISC Corp., Taiwan), which determines the number of steps and the corresponding spin rates and duration. For the protein assay, LC films of E7 were formed on BSA-immobilized substrates by a default three-step spin-coating program of 500 rpm for 10 s, and consecutively 1000 rpm for 10 s and 3000 rpm for 10 s, while for the CA125 immunoassay, LC films of E7 or HDN were formed in a single-step spin-coating procedure at 5000 rpm for 20 s. All spin-coated LC films can be uniformly and stably preserved without shrinkage for several hours at room temperature to allow texture observation or optical measurements to be accomplished.

2.6. Optical LC Texture Observation and Measurement of LC Film Thickness

All measurements were carried out at room temperature at which LCs were in the nematic phase. Optical textures of spin-coated LC films and sandwiched LC cells were observed under crossed polarizers using an Olympus BX51 polarizing optical microscope (POM) in the transmission mode with a 4× objective lens. Microscopic images were taken by an Olympus XC30 digital camera mounted on the microscope with a resolution of 2080 × 1544 pixels and an exposure time of 100 ms. The thickness of the spin-coated LC film was determined by the optical setup as shown in Figure 1. A sample consisting of a DMOAP-coated substrate spin-coated with an LC film was placed between a polarizer and an analyzer whose transmission axes were perpendicular to each other. The incident light source was a He-Ne laser with an emission wavelength of λ = 632.8 nm. In this manner, because the LC film spin-coated on a DMOAP substrate exhibits homeotropic alignment, it can be regarded as a uniaxial crystal film with the optic axis perpendicular to the film plane (i.e., x-y plane), and the correlation between the transmittance (I) of light received by the detector and the angle of light incidence (θ) can be specified as

$$I = \sin^2(2\phi) \sin^2 \left[\frac{\pi n_o d_{LC}}{\lambda} \left(\sqrt{1 - \frac{\sin^2 \theta}{n_e^2}} - \sqrt{1 - \frac{\sin^2 \theta}{n_o^2}} \right) \right] \quad (1)$$

where d_{LC} is the LC film thickness, n_e and n_o are parallel and perpendicular components of the refractive index of LC, respectively, θ is the (polar) angle between the direction of propagation of light (i.e., the z-axis) and the substrate normal (**k**), namely, the unperturbed LC director lying in the x-z plane, and Φ is the angle between the transmission axis of the polarizer (**T**$_p$) and the x-axis [32]. It should be emphasized that the second term in the brackets of Equation (1), estimated based on the law of refraction and the index ellipsoid equation [32], is valid and specific to the phase retardation of a vertically aligned LC film at an arbitrary light incident angle. The value of d_{LC} was deduced from Equation (1) by fitting the measured dependence of I on θ. Note that the conventional interference method to obtain LC film thickness was not applicable in this study because the perpendicular component of the refractive index of LCs (e.g., n_o = 1.52 for E7 and n_o = 1.51 for HDN) is close to that of the glass substrate.

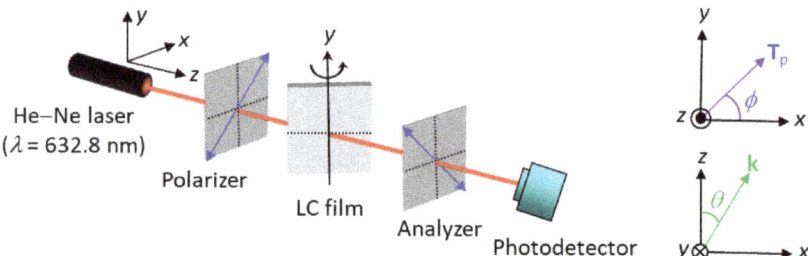

Figure 1. Schematic of the optical setup with crossed polarizers for the measurement of the thickness of LC films spin-coated on a DMOAP-coated glass substrate.

3. Results and Discussion
3.1. LC-Based Single-Substrate Protein Assay with Spin-Coated E7 Film

The LC-based single-substrate protein assay was developed by immobilizing the protein standard BSA on a DMOAP-coated substrate, followed by detection with spin-coated LC film, as depicted in Figure 2. Due to the homeotropic anchoring strength provided by both DMOAP and the air, LC molecules in the semi-free LC film were vertically anchored with their long axes perpendicular to the substrate plane at the LC-DMOAP and LC-air interfaces. A uniformly dark optical LC texture was obtained in the absence of BSA because no phase retardation was experienced when the normally incident polarized light passed through the homeotropically aligned LC film (Figure 2a). When the LC film was spin-coated on a BSA-immobilized DMOAP substrate, LC molecules were reoriented from the homeotropic to the disrupted state. Consequently, the birefringence effect and light scattering caused the appearance of bright but non-uniform optical texture under crossed polarizers (Figure 2b).

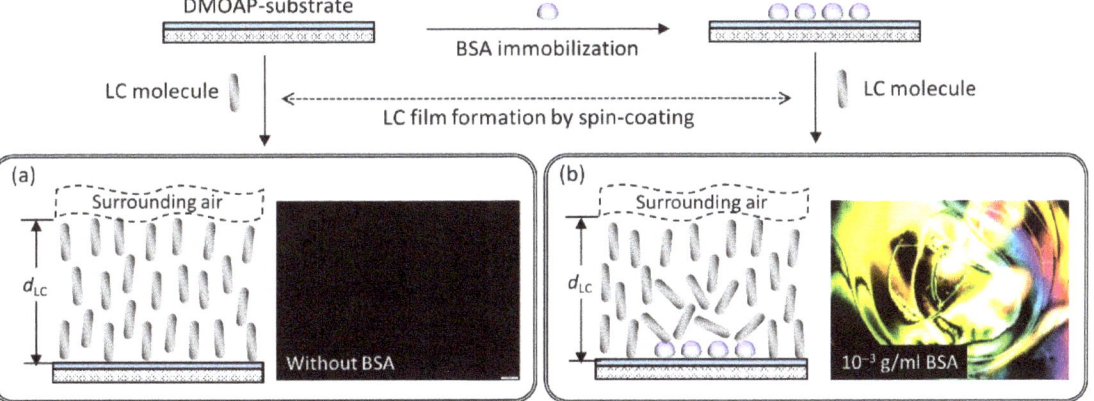

Figure 2. Schematic of the working principle of the LC-based single-substrate protein assay for the detection of BSA with spin-coated LC film on a DMOAP-coated substrate. (**a**) In the absence of BSA, LC molecules are aligned homeotropically, resulting in uniformly dark optical texture. (**b**) In the presence of BSA, LC alignment is disrupted at the LC-glass interface, giving rise to non-uniform and bright optical texture.

Figure 3 compares the optical textures of spin-coated E7 films and sandwiched E7 cells at BSA concentrations (c_{BSA}) between 10^{-7} and 10^{-4} g/mL. Here, the thickness of the E7 film formed by the spin-coater under default conditions described in Section 2.5 was d_{LC} ~5.5 µm, while that in the LC cell was 4.2 ± 0.5 µm. As shown in Figure 3a for spin-coated E7 films on BSA-immobilized DMOAP substrates, the optical texture was uniform with a dark appearance at $c_{BSA} = 10^{-7}$ g/mL but became non-uniform with bright domains in the dark background at $c_{BSA} = 10^{-6}$ g/mL. Increasing the BSA concentration further to $c_{BSA} = 10^{-5}$ and 10^{-4} g/mL resulted in brighter textures than that at $c_{BSA} = 10^{-6}$ g/mL. The limit of detection (LOD) is thus of the order of magnitude of 10^{-6} g/mL, meaning that the amount of immobilized BSA at concentrations lower than 10^{-6} g/mL may be insufficient to weaken the anchoring strength of DMOAP and, in turn, to disrupt the homeotropic orientation of E7. On the other hand, the dark-to-bright optical response to BSA occurred at 10^{-5} g/mL when detected with the sandwiched E7 cell (Figure 3b). At $c_{BSA} = 10^{-6}$ g/mL or lower, a dark texture similar to that in the absence of BSA was observed. These results indicate that using spin-coated LC film instead of the conventional sandwiched LC cell as the sensing platform for protein assay not only simplified the fabrication procedure but enhanced detection sensitivity. This can be explained by the weaker anchoring strength at the LC-air interface compared with that at the LC-DMOAP interface so that LC molecules in the LC film are more easily disrupted in the presence of BSA than those in the LC cell.

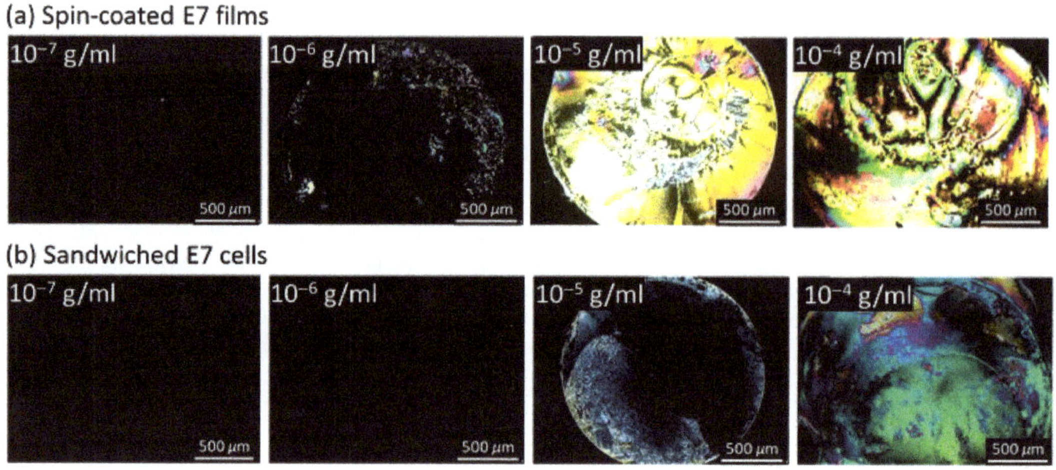

(a) Spin-coated E7 films

(b) Sandwiched E7 cells

Figure 3. Polarized optical textures of E7 in spin-coated films and LC cells in the presence of BSA. The nematic LC E7 was (**a**) spin-coated as a thin film on a DMOAP-coated glass substrate or (**b**) sandwiched between two glass substrates in an LC cell at various BSA concentrations ranging from 10^{-7} to 10^{-4} g/mL. Scale bar, 500 µm.

3.2. Signal Amplification of Single-Substrate Detection through the Control of Film Thickness

The thickness of an LC film can be directly and accurately controlled by adjusting the spin rate. The correlation between the spin rate and LC film thickness was demonstrated with a two-step spin-coating procedure in which the spin rate of the first step (ω_1) was varied from 1000 to 5000 rpm, while that in the second step (ω_2) was fixed at 5000 rpm, with the duration of both steps set at 10 s (Figure 4). The conoscopic image and the average thickness of an E7 film and the corresponding uncertainty were evaluated from five independent sets of experiments to ensure the reproducibility. The homeotropic alignment of the spin-coated E7 film with ω_1 = 1000 rpm, 3000 rpm, or 5000 rpm on a DMOAP substrate was confirmed by the Maltese cross pattern observed in conoscopic images (Figure 4a). The thickness of a spin-coated LC film was determined by measuring optical transmittance at various incident angles of light (θ) with respect to the substrate normal based on the optical setup in Figure 1. As shown in Figure 4b, the transmittance of the LC film increased with increasing θ from 30° to 50°. Because the phase retardation of light passing through an LC film at $\theta < 50°$ is lower than $\pi/2$, thicker LC films resulted in higher transmittance at a fixed θ, according to Equation (1). As a result, by fitting the experimental data in Figure 4b with the equation, the LC film thickness can be deduced as d_{LC} = 4.8 ± 0.3 µm at ω_1 = 1000 rpm, d_{LC} = 4.2 ± 0.2 µm at ω_1 = 3000 rpm, and d_{LC} = 3.4 ± 0.2 µm at ω_1 = 5000 rpm. Note that the uncertainty of the film thickness of ~±0.2 µm is obtained by calculating the standard deviation from five thickness values. This result supports that the LC film can be readily formed on a solid surface with high controllability and reproducibility by using the conventional spin-coating method.

We further investigated the effect of LC film thickness on the LOD for BSA in the proposed LC-based single-substrate protein assay. Here, the brightness of optical images at c_{BSA} = 10^{-8} g/mL (Figure 5a) and 10^{-7} g/mL (Figure 5b) has been artificially increased by 40% to enhance the visibility of bright domains. At a BSA concentration of 10^{-7} or 10^{-6} g/mL, it is demonstrated in Figure 5b and c that the bright domains in the optical texture increased as the film thickness decreased from 4.8 to 3.4 µm. Notably, when BSA concentration was lowered to 10^{-8} g/mL a trace amount of light leakage was still discernible in the optical texture of the 3.4-µm-thick E7 film, while the optical texture was completely dark for the 4.2-µm-thick film (Figure 5a). Good reproducibility of the LOD for BSA detection with spin-coated E7 films was ascertained from the reproducible results at least in four of the five same experiments. For example, in the case of 3.4-μm-thick E7 films spin-coated on BSA-immobilized substrates, consistent results can be obtained from another three sets of experiments at BSA concentrations of 10^{-7}–10^{-9} g/mL, including determinable LOD on the order of 10^{-8} g/mL from dark-to-bright change in the optical image and increasing brightness and bright domains with increasing BSA concentration (Figure 5d). Note that the appearances of repeated optical images were different in general because BSA molecules were free to become immobilized at any place of a given area. The improved LOD for the thinner LC film suggests that the extent of disruption in LC alignment by BSA at the LC-DMOAP interface can be enhanced by reducing the thickness of spin-coated LC films, thus leading to signal amplification. The thickness of LC cells has long been associated with the electro-optical performance of LC display devices. A thin cell exhibits stronger surface interaction and is expected to offer higher image intensity compared with a thick counterpart [7]. Nevertheless, the relevance of LC film thickness to signal amplification in LC-based biosensing has not been implicated until this study, presumably due to the lack of flexibility in controlling the thickness of the LC film. For instance, the LC film formed on an LC–aqueous interface is usually fixed to ~20 μm, conditioned by the thickness of the TEM grid. For manually assembled LC cells, the LC film thickness is determined by the spacer and is typically around 5–10 μm. A smaller cell gap for the LC cell can be achieved but may compromise the uniformity of the sandwiched LC layer and the reproducibility of signal output. Taking advantage of spin-coating, LC film thickness can be directly and easily reduced to a desired smaller value (<5 μm) to enhance the optical response.

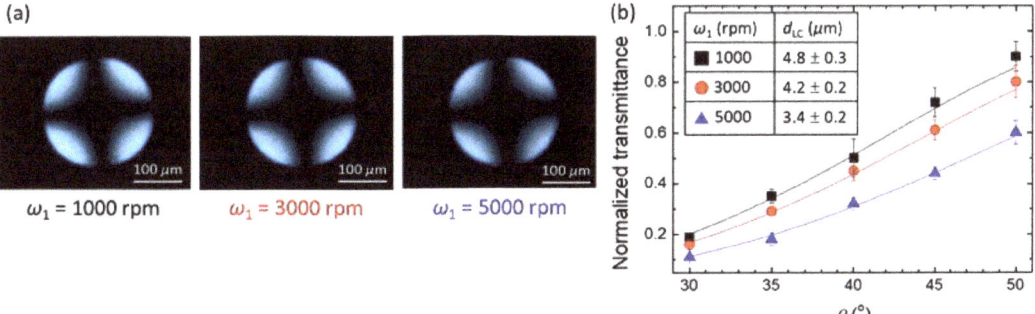

Figure 4. Correlation between spin rate and LC film thickness. The E7 film was spin-coated on DMOAP substrates with a two-step procedure in which the spin rate of the first step ω_1 = 1000, 3000 or 5000 rpm and that of the second step ω_2 = 5000 rpm, with each step lasting for 10 s. (a) Conoscopic images of LC films spread at various ω_1 observed under a POM. Each error bar denotes the standard deviation calculated from the transmission values of five independent measurements. (b) Dependence of transmittance on incident angle of light θ measured with the optical setup in Figure 1. The values of E7 film thickness displayed in the inset of (b) were deduced by fitting the experimental results with Equation (1).

Signal amplification mediated by labeling with gold nanoparticles has been reported in several LC-based biosensors [33–37]. To eliminate the costly and time-consuming procedure of labeling, a number of label-free approaches aimed at enhancing the optical signal and thereby improving detection sensitivity at the LC-solid interface of a sandwiched LC cell were proposed, which include the use of a highly birefringent LC [38,39] and LC-photopolymer composite [12], adjustment of the direction of linearly polarized light for a dye-doped LC [16], modification of the alignment layer by ultraviolet light irradiation [40], and application of a weak electric field to orient LC molecules in a pre-tilted state [41]. Because of the similar working principles between the LC film and LC cell in biodetection at the LC-glass interface, most of the previously reported signal amplification approaches for sandwiched cells would be applicable to the LC film. We have demonstrated that signal amplification can be achieved with both spin-coated film and sandwiched cell of an LC-photopolymer composite [12,31]. In addition, the intensity of the optical response to BSA was enhanced when a 3.4 µm-thick E7 film was spin-coated on DMOAP slides modified with ultraviolet light or when the high-birefringence LC HDN was used instead of E7 to form the sensing LC film, resulting in an improvement in LOD from 10^{-8} to 10^{-9} g/mL BSA (data not shown). In addition to LC–solid interface sensing, a few works have been conducted at the LC–aqueous interface for BSA detection with LOD of ~45 nM using the typical LC-infiltrated TEM grids configuration [42,43]. Notably, it is of perspective to extend the proposed LC-on-a-single-substrate configuration to implementation of protein assay at the LC–aqueous interface because the side open to the air can form an LC–aqueous interface analogous to that of LC-infiltrated TEM grids.

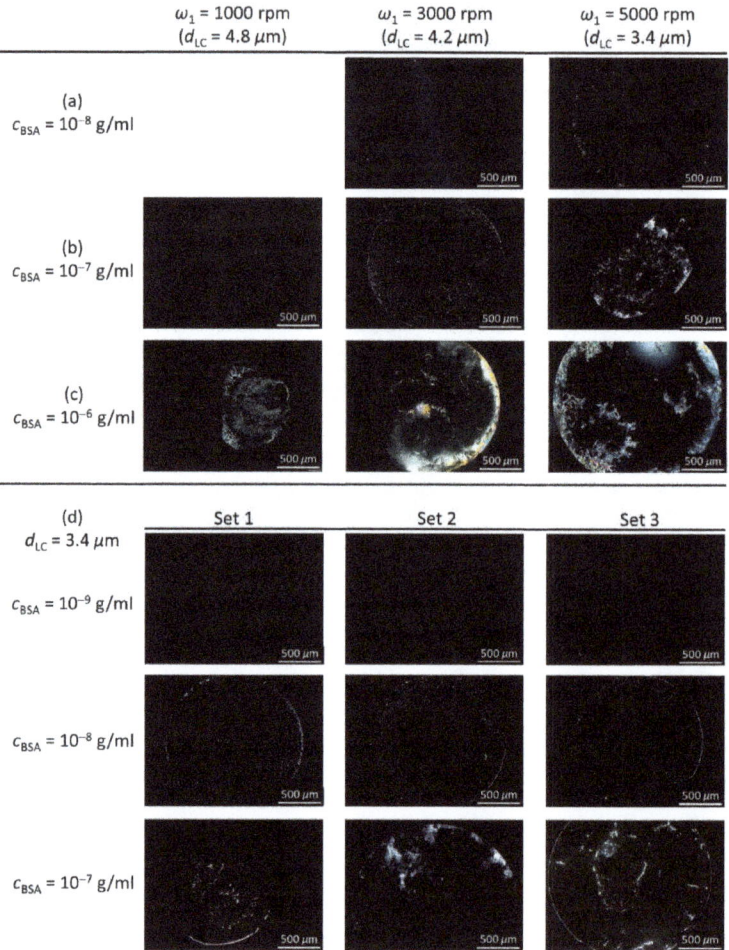

Figure 5. Polarized optical textures of spin-coated E7 films of various thicknesses in the presence of BSA. DMOAP-coated glass substrates with immobilized BSA at concentrations of (a) 10^{-8}, (b) 10^{-7}, and (c) 10^{-6} g/mL were spin-coated with the nematic LC E7 by varying the spin rate of the first step (ω_1) of a two-step spin-coating procedure to form LC films with thicknesses d_{LC} of 4.8, 4.2, and 3.4 µm at ω_1 = 1000, 3000, and 5000 rpm, respectively. (d) Optical textures of three independent sets of experiments for 3.4 µm-thick E7 films in the presence of BSA. Note that the brightness of optical textures in Figure 5a,b,d has been artificially increased by 40% to enhance the visibility. Scale bar, 500 µm.

3.3. LC-Based Single-Substrate Immunoassay for CA125 with Spin-Coated E7 or HDN Film

As illustrated in Figure 6, the sensing principle of an LC-based single-substrate immunoassay depends on the specific binding of the CA125 antigen to anti-CA125 antibody, resulting in the formation of immunocomplexes to induce the reorientation of LC molecules so that CA125 can be detected by the change in optical LC texture under crossed polarizers. The anti-CA125 antibody was first immobilized on a DMOAP-coated substrate as the capture molecule for CA125. To avoid false-positive signals, the amount of immobilized anti-CA125 antibody must be controlled to ensure that the homeotropic orientation of the LC film is not affected by the presence of anti-CA125 antibodies at the LC-DMOAP interface (Figure 6a). As shown in Figure 7, dark optical textures corresponding to homeotropic LC orientation were observed for the spin-coated E7 and HDN films at anti-CA125 antibody concentrations of 10^{-7} and 10^{-8} g/mL, respectively. When the concentration of the anti-CA125 antibody was increased to 10^{-6} g/mL or higher for the E7 film and 10^{-7} g/mL or higher for the HDN film, the optical textures became bright, indicating that the immobilized anti-CA125 antibody alone may reorient LC molecules from the homeotropic to disrupted state, leading to false-positive optical signals in the absence of CA125. As a consequence, in the single-substrate immunoassay based on the E7 and HDN films, the immobilization concentration of the anti-CA125 antibody was limited to 10^{-7} and 10^{-8} g/mL, respectively, which corresponds to the maximal amount of immobilized antibody without background noise. This ensures that the dark-to-bright transition occurring after immunoreaction can be attributed predominantly to CA125 (Figure 6b).

To perform immunodetection of CA125 with the E7 film, 10^{-7} g/mL of the anti-CA125 antibody was immobilized on DMOAP-coated substrates and reacted with different concentrations of CA125, followed by spin-coating of E7. As shown in Figure 8, a red dashed circle is labeled on each micrograph to distinguish the specific binding area (within the circle) immobilized with the anti-CA125 antibody and the nonspecific binding area (outside the circle) where no capture antibody was present. At 10^{-6}-g/mL CA125, the completely dark texture suggests that the amount of the CA125 immunocomplex was too low to induce reorientation of LC molecules (Figure 8a). When the CA125 concentration was increased to 10^{-5} g/mL, a few bright domains appeared in the specific binding area (Figure 8b). At 10^{-4} g/mL CA125, optical response was observed in both the specific and nonspecific binding areas, which connotes that too much CA125 was present such that, in addition to the formation of immunocomplexes through specific binding, excess CA125 also adsorbed nonspecifically to DMOAP in the area without the anti-CA125 antibody (Figure 8c). The LOD of the LC-based immunoassay with spin-coated E7 film was therefore estimated to be of the order of magnitude of 10^{-5} g/mL CA125. By substituting HDN for E7 as the sensing material in the CA125 immunoassay, the optical texture of the HDN film was dark at 10^{-9} g/mL CA125 (Figure 9a), but its brightness gradually increased with CA125 concentration from 10^{-8} to 10^{-5} g/mL (Figure 9b–e). The lower LOD of the HDN film, which was of the order of magnitude of 10^{-8} g/mL CA125, achieved at an anti-CA125 antibody concentration (10^{-8} g/mL) an order of magnitude lower than that for the E7 film can be explained by the higher birefringence of HDN compared with E7. It is ensured that the above-mentioned results specific to immunodetection of CA125 with spin-coated LC films were reproducible at least in three of five sets of experiments, for example, similar optical textures in sets 1–3 for spin-coated HDN LC films in the presence of CA125 as shown in Figure 9. As a comparison, the LOD of a CA125 immunoassay based on sandwiched HDN LC cells was 10^{-8} g/mL CA125 at an anti-CA125 antibody immobilization concentration of 10^{-7} g/mL [38]. It is therefore concluded from the comparable LOD attained with relatively less capture antibody by the LC film that the performance of LC-based optical biosensors can be improved by replacing the sandwiched LC cell with a spin-coated LC film in the single-substrate biosensing platform.

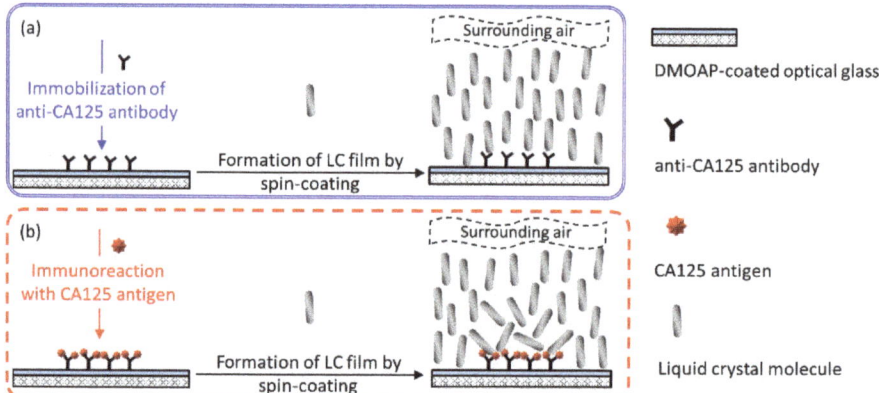

Figure 6. Schematics of the working principle of the LC-based single-substrate immunoassay for the detection of CA125 with spin-coated LC film on a DMOAP-coated substrate. (**a**) In the absence of CA125, LC molecules are aligned homeotropically on immobilized anti-CA125 antibody, whose amount is adjusted to a maximum without disturbing the orientation of LCs. (**b**) In the presence of CA125, alignment of LC molecules is disrupted due to the formation of the CA125 immunocomplex atop the DMOAP aligning layer.

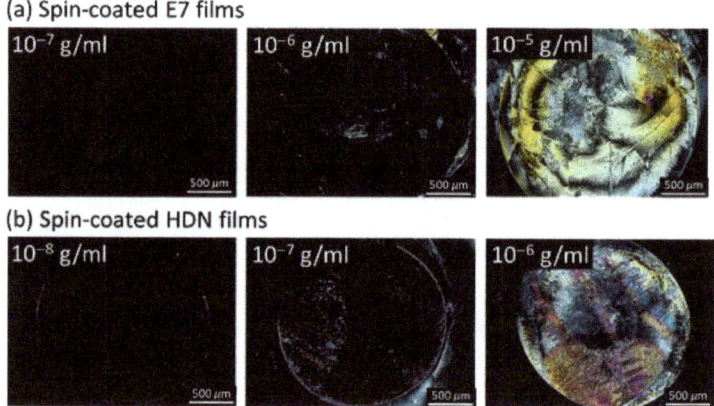

Figure 7. Polarized optical textures of spin-coated E7 and HDN films in the presence of the anti-CA125 antibody. The nematic LCs (**a**) E7 and (**b**) HDN were spin-coated on DMOAP-coated substrates immobilized with the anti-CA125 antibody at concentrations ranging from 10^{-8} to 10^{-5} g/mL. Scale bar, 500 μm.

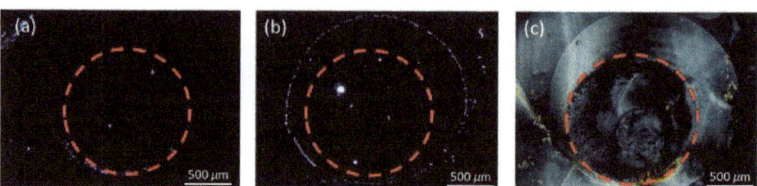

Figure 8. Polarized optical textures of spin-coated E7 films in the presence of the CA125 protein. The nematic LC E7 was spin-coated on DMOAP-coated substrates immobilized with 10^{-7} g/mL anti-CA125 antibody and reacted with CA125 at concentrations of (**a**) 10^{-6}, (**b**) 10^{-5}, and (**c**) 10^{-4} g/mL. Each red dashed circle represents the area within which the anti-CA125 antibody was immobilized. Scale bar, 500 μm.

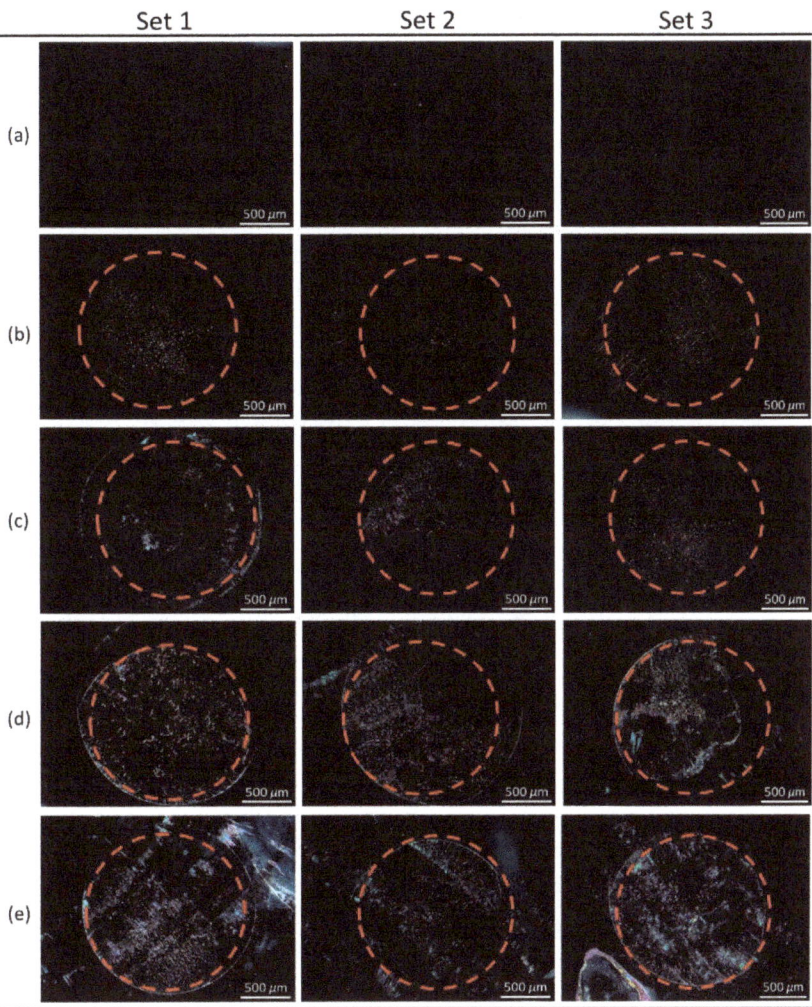

Figure 9. Polarized optical textures of spin-coated HDN films in the presence of the CA125 protein. The nematic LC HDN was spin-coated on DMOAP-coated substrates immobilized with 10^{-8} g/mL anti-CA125 antibody and reacted with CA125 at concentrations of (**a**) 10^{-9}, (**b**) 10^{-8}, (**c**) 10^{-7}, (**d**) 10^{-6}, and (**e**) 10^{-5} g/mL. Each red dashed circle represents the area within which the anti-CA125 antibody was immobilized. Sets 1–3 represent results of three independent experiments. Scale bar, 500 μm.

4. Conclusions

In summary, a label-free single-substrate biosensor based on a spin-coated LC film was established in this study to achieve lower limit in protein detection and immunoassay. The semi-free LC film was supported by the asymmetric homeotropic anchoring strengths at the LC-air and LC-DMOAP interfaces to align LC molecules vertically with their long axes parallel to the substrate normal. Disruption of the ordered orientation of LCs by the presence of biomolecules at the LC-DMOAP interface leads to dark-to-bright transition of the optical LC texture, giving rise to a high signal-to-background detection signal. Compared to the LC cell, the thickness of the spin-coated LC film is more easily reduced to a thickness smaller than 5 μm by adjusting the spin rate to improve detection sensitivity. When the nematic LC E7 was used in the detection of BSA, the LOD of a 4.2-μm-thick spin-coated E7 film was estimated to be 10^{-7}-g/mL BSA, which was two orders of magnitude lower than that of a 4.2-μm-thick sandwiched E7 film in an LC cell. Moreover, by reducing the E7 film thickness to 3.4 μm, the LOD can be further improved to 10^{-8}-g/mL BSA. The potential for clinical application of the LC-based single-substrate biosensor was demonstrated with an immunoassay for the cancer biomarker CA125, in which the LOD was determined as 10^{-5}-g/mL CA125 at an anti-CA125 antibody concentration of 10^{-7} g/mL for the E7 film. Substituting HDN, a nematic LC of higher birefringence, for E7 as the sensing medium resulted in a lower LOD of 10^{-8}-g/mL CA125 with 10^{-8}-g/mL anti-CA125 antibody immobilized. It is evident from the results of this study that, in addition to the ease of preparation, single-substrate detection with spin-coated LC film offered a new means of signal amplification by reducing film thickness to improve LOD and detection sensitivity. With a wide variety of LCs available in the industries, combined with numerous surface modification and patterning techniques to stabilize the LC film, new possibilities are revealed for the development of more advanced LC-based single-substrate biosensing technologies to extend their practical application.

Author Contributions: Conceptualization, P.-C.W. and W.L.; Methodology, P.-C.W. and C.-P.P.; Software, C.-P.P.; Validation, P.-C.W., M.-J.L. and W.L.; Formal Analysis, P.-C.W. and C.-P.P.; Investigation, C.-P.P. and P.-C.W.; Resources, M.-J.L. and W.L.; Data Curation, P.-C.W. and C.-P.P.; Writing—Original Draft Preparation, P.-C.W.; Writing—Review and Editing, M.-J.L. and W.L.; Visualization, P.-C.W. and C.-P.P.; Supervision, M.-J.L. and W.L.; Project Administration, M.-J.L. and W.L.; Funding Acquisition, M.-J.L. and W.L. All authors have read and agreed to the published version of the manuscript.

Funding: This work was financially supported by the Ministry of Science and Technology, Taiwan, under grant Nos. 107-2112-M-009-012-MY3, 109-2320-B-309-001, 110-2112-M-A49-023, and 110-2320-B-309-001.

Institutional Review Board Statement: Not applicable.

Informed Consent Statement: Not applicable.

Data Availability Statement: The authors confirm that the data supporting the findings of this study are available within the article.

Conflicts of Interest: The authors declare no conflict of interest.

References

1. Gupta, V.K.; Skaife, J.J.; Dubrovsky, T.B.; Abbott, N.L. Optical amplification of ligand-receptor binding using liquid crystals. *Science* **1998**, *279*, 2077–2080. [CrossRef] [PubMed]
2. Miller, D.S.; Carlton, R.J.; Mushenheim, P.C.; Abbott, N.L. Introduction to optical methods for characterizing liquid crystals at interfaces. *Langmuir* **2013**, *29*, 3154–3169. [CrossRef]
3. Luan, C.; Luan, H.; Luo, D. Application and technique of liquid crystal-based biosensors. *Micromachines* **2020**, *11*, 176. [CrossRef]
4. Wang, Z.; Xu, T.; Noel, A.; Chen, Y.-C.; Liu, T. Applications of liquid crystals in biosensing. *Soft Matter* **2021**, *17*, 4675–4702. [CrossRef]
5. Carlton, R.J.; Hunter, J.T.; Miller, D.S.; Abbasi, R.; Mushenheim, P.C.; Tan, L.N.; Abbott, N.L. Chemical and biological sensing using liquid crystals. *Liq. Cryst. Rev.* **2013**, *1*, 29–51. [CrossRef]
6. Hussain, Z.; Qazi, F.; Ahmed, M.I.; Usman, A.; Riaz, A.; Abbasi, A.D. Liquid crystals based sensing platform-technological aspects. *Biosens. Bioelectron.* **2016**, *85*, 110–127. [CrossRef]
7. Prakash, J.; Parveen, A.; Mishra, Y.K.; Kaushik, A.K. Nanotechnology-assisted liquid crystals-based biosensors: Towards fundamental to advanced applications. *Biosens. Bioelectron.* **2020**, 112562. [CrossRef]
8. Popov, N.; Honaker, L.W.; Popova, M.; Usol'tseva, N.; Mann, E.K.; Jákli, A.; Popov, P. Thermotropic liquid crystal-assisted chemical and biological sensors. *Materials* **2018**, *11*, 20. [CrossRef] [PubMed]
9. Popov, P.; Mann, E.K.; Jákli, A. Thermotropic liquid crystal films for biosensors and beyond. *J. Mater. Chem. B* **2017**, *5*, 5061–5078. [CrossRef] [PubMed]
10. Lin, C.-H.; Lee, M.-J.; Lee, W. Bovine serum albumin detection and quantitation based on capacitance measurements of liquid crystals. *Appl. Phys. Lett.* **2016**, *109*, 093703. [CrossRef]
11. Lin, C.-M.; Wu, P.-C.; Lee, M.-J.; Lee, W. Label-free protein quantitation by dielectric spectroscopy of dual-frequency liquid crystal. *Sens. Actuators B Chem.* **2019**, *282*, 158–163. [CrossRef]

12. Shaban, H.; Yen, S.C.; Lee, M.J.; Lee, W. Signal amplification in an optical and dielectric biosensor employing liquid crystal-photopolymer composite as the sensing medium. *Biosensors* **2021**, *11*, 81. [CrossRef]
13. Hsiao, Y.-C.; Sung, Y.-C.; Lee, M.-J.; Lee, W. Highly sensitive color-indicating and quantitative biosensor based on cholesteric liquid crystal. *Biomed. Opt. Express* **2015**, *6*, 5033–5038. [CrossRef]
14. Lee, M.-J.; Chang, C.-H.; Lee, W. Label-free protein sensing by employing blue phase liquid crystal. *Biomed. Opt. Express* **2017**, *8*, 1712–1720. [CrossRef] [PubMed]
15. Lee, M.-J.; Pai, C.-P.; Wu, P.-C.; Lee, W. Label-free single-substrate quantitative protein assay based on optical characteristics of cholesteric liquid crystals. *J. Mol. Liq.* **2021**, *331*, 115756. [CrossRef]
16. Chiang, Y.-L.; Lee, M.-J.; Lee, W. Enhancing detection sensitivity in quantitative protein detection based on dye-doped liquid crystals. *Dyes Pigment.* **2018**, *157*, 117–122. [CrossRef]
17. Wu, P.-C.; Karn, A.; Lee, M.-J.; Lee, W.; Chen, C.-Y. Dye-liquid-crystal-based biosensing for quantitative protein assay. *Dyes Pigment.* **2018**, *150*, 73–78. [CrossRef]
18. Brake, J.M.; Mezera, A.D.; Abbott, N.L. Effect of surfactant structure on the orientation of liquid crystals at aqueous—liquid crystal interfaces. *Langmuir* **2003**, *19*, 6436–6442. [CrossRef]
19. Hussain, Z.; Zafiu, C.; Küpcü, S.; Pivetta, L.; Hollfelder, N.; Masutani, A.; Kilickiran, P.; Sinner, E.-K. Liquid crystal based sensors monitoring lipase activity: A new rapid and sensitive method for cytotoxicity assays. *Biosens. Bioelectron.* **2014**, *56*, 210–216. [CrossRef]
20. Hartono, D.; Bi, X.; Yang, K.L.; Yung, L.Y.L. An air-supported liquid crystal system for real-time and label-free characterization of phospholipases and their inhibitors. *Adv. Funct. Mater.* **2008**, *18*, 2938–2945. [CrossRef]
21. Popov, P.; Honaker, L.W.; Kooijman, E.E.; Mann, E.K.; Jákli, A.I. A liquid crystal biosensor for specific detection of antigens. *Sens. Bio-Sens. Res.* **2016**, *8*, 31–35. [CrossRef]
22. Xue, C.-Y.; Yang, K.-L. Dark-to-bright optical responses of liquid crystals supported on solid surfaces decorated with proteins. *Langmuir* **2008**, *24*, 563–567. [CrossRef] [PubMed]
23. Zhu, Q.; Yang, K.-L. Microfluidic immunoassay with plug-in liquid crystal for optical detection of antibody. *Anal. Chim. Acta* **2015**, *853*, 696–701. [CrossRef] [PubMed]
24. Fan, Y.-J.; Chen, F.-L.; Liou, J.-C.; Huang, Y.-W.; Chen, C.-H.; Hong, Z.-Y.; Lin, J.-D.; Hsiao, Y.-C. Label-free multi-microfluidic immunoassays with liquid crystals on polydimethylsiloxane biosensing chips. *Polymers* **2020**, *12*, 395. [CrossRef]
25. Huang, J.-W.; Hisamoto, H.; Chen, C.-H. Quantitative analysis of liquid crystal-based immunoassay using rectangular capillaries as sensing platform. *Opt. Express* **2019**, *27*, 17080–17090. [CrossRef]
26. Liu, D.; Jang, C.-H. A new strategy for imaging urease activity using liquid crystal droplet patterns formed on solid surfaces. *Sens. Actuators B Chem.* **2014**, *193*, 770–773. [CrossRef]
27. Han, G.-R.; Jang, C.-H. Detection of heavy-metal ions using liquid crystal droplet patterns modulated by interaction between negatively charged carboxylate and heavy-metal cations. *Talanta* **2014**, *128*, 44–50. [CrossRef]
28. Dhara, P.; Mukherjee, R. Phase transition and dewetting of a 5CB liquid crystal thin film on a topographically patterned substrate. *RSC Adv.* **2019**, *9*, 21685–21694. [CrossRef]
29. Dhara, P.; Bhandaru, N.; Das, A.; Mukherjee, R. Transition from spin dewetting to continuous film in spin coating of liquid crystal 5CB. *Sci. Rep.* **2018**, *8*, 7169. [CrossRef]
30. Ma, L.-L.; Duan, W.; Tang, M.-J.; Chen, L.-J.; Liang, X.; Lu, Y.-Q.; Hu, W. Light-driven rotation and pitch tuning of self-organized cholesteric gratings formed in a semi-free film. *Polymers* **2017**, *9*, 295. [CrossRef]
31. Lee, M.J.; Duan, F.F.; Wu, P.C.; Lee, W. Liquid crystal-photopolymer composite films for label-free single-substrate protein quantitation and immunoassay. *Biomed. Opt. Express* **2020**, *11*, 4915–4927. [CrossRef]
32. Cheng, H.-C.; Yan, J.; Ishinabe, T.; Sugiura, N.; Liu, C.-Y.; Huang, T.-H.; Tsai, C.-Y.; Lin, C.-H.; Wu, S.-T. Blue-phase liquid crystal displays with vertical field switching. *J. Disp. Technol.* **2012**, *8*, 98–103. [CrossRef]
33. Liao, S.; Qiao, Y.; Han, W.; Xie, Z.; Wu, Z.; Shen, G.; Yu, R. Acetylcholinesterase liquid crystal biosensor based on modulated growth of gold nanoparticles for amplified detection of acetylcholine and inhibitor. *Anal. Chem.* **2012**, *84*, 45–49. [CrossRef] [PubMed]
34. Yang, S.; Liu, Y.; Tan, H.; Wu, C.; Wu, Z.; Shen, G.; Yu, R. Gold nanoparticle based signal enhancement liquid crystal biosensors for DNA hybridization assays. *Chem. Commun.* **2012**, *48*, 2861–2863. [CrossRef] [PubMed]
35. Zhao, D.; Peng, Y.; Xu, L.; Zhou, W.; Wang, Q.; Guo, L. Liquid-crystal biosensor based on nickel-nanosphere-induced homeotropic alignment for the amplified detection of thrombin. *ACS Appl. Mater. Interfaces* **2015**, *7*, 23418–23422. [CrossRef]
36. Wang, Y.; Wang, B.; Xiong, X.; Deng, S. Gold nanoparticle-based signal enhancement of an aptasensor for ractopamine using liquid crystal based optical imaging. *Microchim. Acta* **2019**, *186*, 697. [CrossRef] [PubMed]
37. Nandi, R.; Loitongbam, L.; De, J.; Jain, V.; Pal, S.K. Gold nanoparticle-mediated signal amplification of liquid crystal biosensors for dopamine. *Analyst* **2019**, *144*, 1110–1114. [CrossRef]
38. Su, H.-W.; Lee, Y.-H.; Lee, M.-J.; Hsu, Y.-C.; Lee, W. Label-free immunodetection of the cancer biomarker CA125 using high-Δn liquid crystals. *J. Biomed. Opt.* **2014**, *19*, 077006. [CrossRef]
39. Sun, S.-H.; Lee, M.-J.; Lee, Y.-H.; Lee, W.; Song, X.; Chen, C.-Y. Immunoassays for the cancer biomarker CA125 based on a large-birefringence nematic liquid-crystal mixture. *Biomed. Opt. Express* **2015**, *6*, 245–256. [CrossRef]

40. Su, H.-W.; Lee, M.-J.; Lee, W. Surface modification of alignment layer by ultraviolet irradiation to dramatically improve the detection limit of liquid-crystal-based immunoassay for the cancer biomarker CA125. *J. Biomed. Opt.* **2015**, *20*, 057004. [CrossRef]
41. Hsu, W.-L.; Lee, M.-J.; Lee, W. Electric-field-assisted signal amplification for label-free liquid-crystal-based detection of biomolecules. *Biomed. Opt. Express* **2019**, *10*, 4987–4998. [CrossRef] [PubMed]
42. Omer, M.; Park, S.-Y. Preparation of QP4VP-b-LCP liquid crystal block copolymer and its application as a biosensor. *Anal. Bioanal. Chem.* **2014**, *406*, 5369–5378. [CrossRef] [PubMed]
43. Cui, X.; Ren, L.; Shan, Y.; Wang, X.; Yang, Z.; Li, C.; Xu, J.; Ma, B. Smartphone-based rapid quantification of viable bacteria by single-cell microdroplet turbidity imaging. *Analyst* **2018**, *143*, 3309–3316. [CrossRef] [PubMed]

Article

Millimeter-Wave-Based Spoof Localized Surface Plasmonic Resonator for Sensing Glucose Concentration

Yelim Kim, Ahmed Salim and Sungjoon Lim *

School of Electrical and Electronics Engineering, Chung-Ang University, Seoul 06974, Korea; kyelim6915@naver.com (Y.K.); ahmedsalim789@gmail.com (A.S.)
* Correspondence: sungjoon@cau.ac.kr

Abstract: Glucose-monitoring sensors are necessary and have been extensively studied to prevent and control health problems caused by diabetes. Spoof localized surface plasmon (LSP) resonance sensors have been investigated for chemical sensing and biosensing. A spoof LSP has similar characteristics to an LSP in the microwave or terahertz frequency range but with certain advantages, such as a high-quality factor and improved sensitivity. In general, microwave spoof LSP resonator-based glucose sensors have been studied. In this study, a millimeter-wave-based spoof surface plasmonic resonator sensor is designed to measure glucose concentrations. The millimeter-wave-based sensor has a smaller chip size and higher sensitivity than microwave-frequency sensors. Therefore, the microfluidic channel was designed to be reusable and able to operate with a small sample volume. For alignment, a polydimethylsiloxane channel was simultaneously fabricated using a multilayer bonding film to attach the upper side of the pattern, which is concentrated in the electromagnetic field. This real-time sensor detects the glucose concentration via changes in the S11 parameter and operates at 28 GHz with an average sensitivity of 0.015669 dB/(mg/dL) within the 0–300 mg/dL range. The minimum detectable concentration and the distinguishable signal are 1 mg/dL and 0.015669 dB, respectively, from a 3.4 µL sample. The reusability and reproducibility were assessed through replicates.

Keywords: spoof localized surface plasmon polariton; sensor; glucose solution; millimeter wave; metamaterial

Citation: Kim, Y.; Salim, A.; Lim, S. Millimeter-Wave-Based Spoof Localized Surface Plasmonic Resonator for Sensing Glucose Concentration. *Biosensors* **2021**, *11*, 358. https://doi.org/10.3390/bios11100358

Received: 25 August 2021
Accepted: 25 September 2021
Published: 28 September 2021

Publisher's Note: MDPI stays neutral with regard to jurisdictional claims in published maps and institutional affiliations.

Copyright: © 2021 by the authors. Licensee MDPI, Basel, Switzerland. This article is an open access article distributed under the terms and conditions of the Creative Commons Attribution (CC BY) license (https://creativecommons.org/licenses/by/4.0/).

1. Introduction

Diabetes is a serious disease that currently affects the health of large populations worldwide and causes various health and lifestyle complications. Globally, the number of diabetic patients currently accounts for >8.5% of the adult population, and the prevalence has been steadily increasing. Diabetes is caused by inadequate absorption of glucose from the blood, mainly due to problems with insulin production. Hence, blood glucose concentration monitoring is necessary to prevent and control diabetes and related complications. Thus, technologies sensing blood glucose concentration are attracting much attention in the medical field. Sensing techniques, such as thermal, optical, mechanical, and microwave-based methods, have been proposed. Glucose sensors using any of these technologies should be small in size and be able to operate multiple times and on small samples, with high sensitivity, accuracy, and resolution [1–6].

In general, electrochemical or optical sensors are used for measuring glucose concentration. For example, electrochemical enzyme-based sensors involving finger pricking is widely used. Although there are few commercial microwave-based sensors, these sensors are attracting much attention due to their advantages, such as non-invasiveness, low cost, and easy fabrication. In microwave-based glucose sensors, an epsilon negative unit-cell resonator, complementary electric LC resonator, or passive components are used [7–10].

Localized surface plasmon (LSP) is defined as the confinement of a surface plasmon (SP) to nanoparticle size. SP refers to electromagnetic field (E-field) propagation along the interface between a metal and a dielectric material at the optical frequency, and LSP is the oscillations of free electrons on metallic surfaces. LSP-based methods are used in many applications, such as lenses, waveguides, and solar cells. In sensor applications, LSP sensors typically have high sensitivity because of their confined mode profiles and near-field enhancements [11–13]. In addition, various methods have been studied for optical LSP-based glucose sensing.

However, these methods are only used at the optical frequency. The spoof LSP-based sensor has also been studied because of its merits, such as a high-quality factor (Q-factor) and high sensitivity. This method has been developed from periodic holes made by ultrathin corrugated metallic disks. In addition, corrugated spoof LSP resonators have been developed in several shapes to improve the Q-factor. Therefore, we designed a millimeter-wave-based spoof LSP resonator sensor to achieve small physical size and sample volume, as well as a high Q-factor and sensitivity level. The structure used to construct these spoof LSPs is called the plasmonic metamaterial. The metamaterial is an artificial material that has negative indices of refraction and is generally used in applications such as antennas, absorbers, and lenses [14–20]. In general, the corrugated structure used in spoof LSP resonator-based sensors operates at the microwave frequency.

In this study, a spoof LSP resonator was used to design a glucose sensor that operates at millimeter-wave frequency. Millimeter-wave-based sensors have the merits of high data transmission rates for communication, enhanced security, and reduced interference while supporting miniaturized sensor sizes [21,22]. In addition, this sensor uses a microfluidic channel fabricated from polydimethylsiloxane (PDMS), and channels are connected by using a multilayer bonding film. The designed PDMS channel reduces the sample volume and increases the detection sensitivity. The channel through which the glucose solution is injected is considered the loading position, where the E-field of the sensor is concentrated. Further, the channel structure reduces the effects of air bubbles. The multilayer bonding film reduces any remaining sample noise, and our design confers a high Q-factor to the spoof LSP resonator. Accordingly, the sensor presented here has a small size, high Q-factor, and high sensitivity, and can operate on small volumes of samples.

We fabricated sensors by using either of two different designs of microfluidic channels. The resulting sensors could detect concentrations in the range of 0–300 mg/dL. The channels were designed and measured nine samples to assess the sensing performance of the sensor. The sensors detect differences in glucose concentration according to changes in the reflection coefficients; the magnitude of the reflection coefficient increases with the glucose concentration. In addition, sensor sensitivity, reusability, and reproducibility were evaluated. The average sensitivity was 0.015669 dB with a 3.4 µL sample volume, and the sensors could be used up to 60 times. Accordingly, their reproducibility was approximately 0.3%. The results obtained using the PDMS channel were compared with the results obtained from sensors based on spoof SPs or LSPs as well as other state-of-the-art glucose sensors. The proposed sensor was observed to have high sensitivity for a miniaturized device that can operate on small volumes of samples. The sensor also has a low detection limit of 1 mg/dL.

2. Materials and Methods

2.1. Preparation of the Materials and Glucose-Solution Sample

The PDMS and bonding film used to fabricate the microfluidic channels were manufactured by Shielding Solutions Ltd., Braintree, UK, and Adhesives Research, Glen Rock, PA, USA, respectively. D-(+)-Glucose powder and deionized (DI) water (pH 6.4) were purchased from Sigma Aldrich. The DI water was produced via reversed osmosis. Glucose-solution samples were prepared in-house via mixing at 40–45 kHz.

2.2. Complex Permittivity of the Solution Samples

DI water and glucose solutions were prepared to investigate the detection performances of the designed glucose sensors. For the measurements, glucose-solution samples with concentrations of 0–300 mg/dL were prepared. The complex relative permittivity was measured using DI water and 10 and 20 mg/dL glucose, and the temperature of the liquid was maintained at 27.8 °C. Figure 1 shows the measured complex permittivity of the DI water and glucose solutions from 15 to 25 GHz. The proposed sensor has an operating frequency range up to 40 GHz; however, owing to the accuracy limits of the machine, permittivity was measured only until 25 GHz. To measure these electromagnetic properties of the sample solutions, we used Keysight N1501A and 8510C equipment. These properties can also be measured via various techniques, such as using a resonator, a coaxial probe, and a cavity. ε' and ε'' are the real and imaginary parts of the complex relative permittivity. Figure 1a shows the measured complex relative permittivity of the DI water. The real part ε' decreased from 50.7871 to 32.4382, and the imaginary part ε'' increased from 35.4508 to approximately 20 GHz and then decreased to 35.4901.

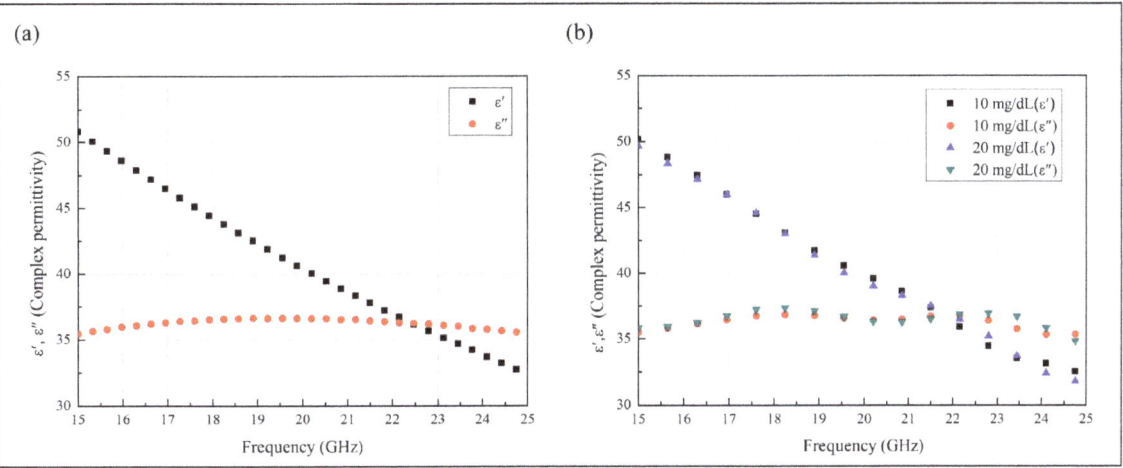

Figure 1. Complex relative permittivity ε' and ε'' of (**a**) deionized water and (**b**) 10 and 20 mg/dL glucose, all measured at 27.8 °C.

Figure 1b shows the measured permittivity of 10 and 20 mg/dL glucose. With a 10 mg/dL concentration, ε' decreased from 50.1766 to 31.7046, and ε'' increased from 35.5241 to approximately 20 GHz and decreased thereafter to 35.1749. With a 20 mg/dL concentration, ε' decreased from 49.6697 to 31.292, and ε'' increased from 35.8409 to approximately 20 GHz and decreased thereafter to 36.8368. When the glucose concentration was increased by 10 mg/dL, the average values of ε' and ε'' increased by 0.49 and decreased by 0.13, respectively.

The complex relative permittivity and loss tangent were defined by the measured values of ε' and ε''. Equation (1) represents the complex relative permittivity ε_c:

$$\varepsilon_c = \varepsilon' + j\varepsilon''. \tag{1}$$

Further, the tangent loss $tan\delta$ can be defined as the ratio of the real part to the imaginary part of the complex relative permittivity, as follows:

$$tan\delta = \frac{\varepsilon'}{\varepsilon''}. \tag{2}$$

These results show the dielectric properties of the prepared DI water and glucose-solution samples. These dielectric properties depend on the frequency used. The values calculated using Equations (1) and (2) represent the changes in the glucose concentration, which affect the dielectric constant and loss tangent. With increases in the glucose concentration, the dielectric constant decreases, while the loss tangent increases. Moreover, the temperature changes also affect the dielectric constant and loss tangent of glucose solution. Thus, the samples were maintained at a constant temperature during the measurements for accuracy [23–26].

2.3. Sensor Design Based on the Spoof LSP Resonator

Figure 2 shows the schematic of the proposed sensor and microfluidic channels with a multilayer bonding film. The proposed sensor consists of two layers and each layer is fabricated using the same Rogers Duroid 5880 substrate. The thickness of the substrate is 0.25 mm and that of the attached copper is 0.018 mm; the dielectric constant ε_r and loss tangent $tan\delta$ of the substrate are 2.2 and 0.0009, respectively. Figure 2a shows the top view of the designed sensor, with the parameters of the top layer pattern. The top substrate has no ground and only the top pattern. The length from the center of the spoof LSP resonator pattern to the edge of the substrate is 4.9 mm. The pattern has a ring-shaped resonator with periodic grooves and a higher Q-factor than the conventional spoof LSP resonator. This modified resonator has the merits of restraining high-order modes and enhancing the fundamental mode. The width S_w and length S_h of the top substrate are 20 and 14.9 mm, respectively. The diameter G_r and width G_w of the interior of the ring are 1.4 and 0.1 mm, respectively. The parameter V_r of the bottom substrate is the hole for connecting the K-band connectors and has a diameter of 1.98 mm. The bottom substrate consists of the ground and a 50 Ω microstrip line. The microstrip line has a circular end for reducing the reflected waves. The circular shape increases the efficiency of the transfer of the E-field to the pattern. The bottom substrate has the same width as the top substrate but a different length S_l (20 mm) [27]. Figure 2b shows the 3D view of each layer of the sensor, locations of the bonding films, and the microfluidic channel. The width of the microstrip line, M_w, is 0.7 mm and the length from the center of the circle M_l is 8.8 mm; the diameter of the circle, M_r, is 1.4 mm. The bonding film connects the top and bottom substrates and has the same size as the top substrate. The bonding film between the substrates is only a 0.12 mm-thick single adhesion layer. The PDMS channels are arranged at the center of the pattern; in the sensor design, the ground, microstrip line, and pattern were fabricated using copper, which has an electrical conductivity of $\sigma = 5.8 \times 10^{-7}$ Sm^{-1}.

Figure 2c,d shows the microfluidic channels made of PDMS and the bonding film layers constructed via the laser-cutting fabrication method. The thickness of the PDMS is 1 mm, and the bonding film is 0.12 mm thick at the adhesion region and 0.02 mm thick at the film region. The dielectric constant ε_r and loss tangent $tan\delta$ of the PDMS are 2.7 and 0.035, respectively. The adhesion and film regions have ε_r values of 2.35 and 2, respectively, with the same $tan\delta$ value of 0.002. Both the PDMS walls and channels were constructed using the same fabrication technique. The multilayer bonding film is composed of three layers, and the first adhesion layer with the same structure as the microfluidic channel is produced simultaneously with the PDMS using a laser. The bonding film, which consists of film and adhesion layers attached to the PDMS, was cut using a laser, and only the film layer was removed. Then, the film layer was attached to the side of the adhesion region of PDMS. This fabrication method was considered to prevent the misalignment between the channel and the bonding film and the remaining solution. When the bonding film is misaligned, the solution may leak, and small liquid bubbles can attach to the corners. Removal of the remaining sample solution is therefore necessary for accurate results. In addition, the middle layer film region allows clearer solution sample removal compared to the adhesion layer. Because the film region has less surface roughness and less adhesion to liquids than the conventional single-layer bonding film, which has an adhesion region, less sample remains in this method than the conventional method. Further, due to the high sensitivity of the sensor, the results are affected by the alignment of the microfluidic channel with the pattern. Alignment lines marked on the top and bottom layers are therefore used to reduce such positioning problems.

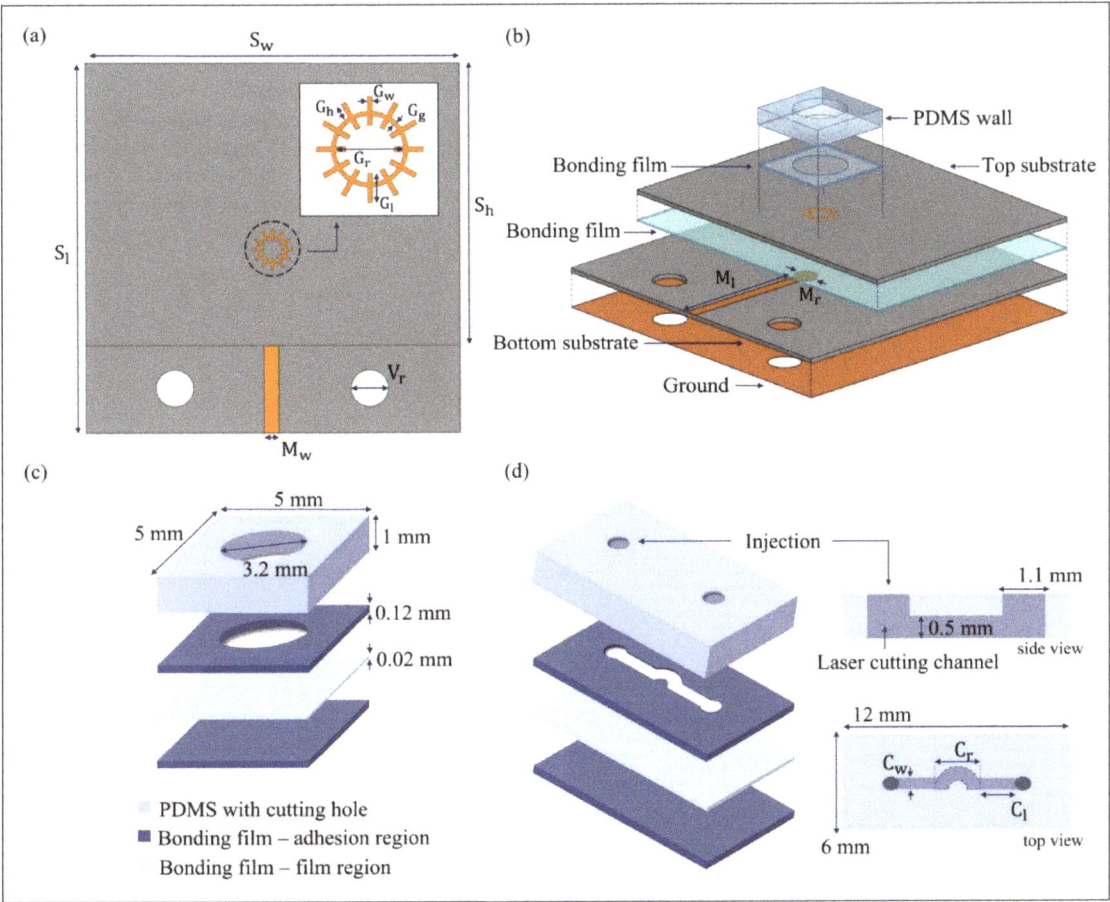

Figure 2. Schematic of the proposed sensor and PDMS channel. (**a**) Top view of the sensor and magnified design of the spoof LSP resonator. (**b**) Layer design of the sensor with microfluidic channels and their alignments. (**c**) PDMS wall and bonding film layer. (**d**) Detailed PDMS-channel design showing the bonding film layer.

Figure 2c shows the design of the PDMS wall, which is 5 × 5 × 1 mm in size. The diameter of the wall is 3.2 mm; hence, the inside volume of the wall is 8.04 mm^3. Thus, an 8 µL glucose solution is injected using a micropipette. Figure 2d shows the design of the PDMS channel and detailed parameters. The channel width and length are 12 and 6 mm, respectively. The diameter of the injection hole is 1.1 mm, and the internal channel thickness is 0.5 mm. The width of the channel, C_w, is 0.8 mm, including the curved side. The channel length from the edge of the curved outside diameter to the surface of the injection hole, C_l, is 1.6 mm. The diameter of the curved channel area, C_r, is 2.4 mm.

The PDMS channel is designed to increase sensitivity. Figure 3 shows the simulated E-field concentration of the sensor at the resonant frequency of 37.02 GHz. The figure shows the magnitude of the E-field vector in the +z direction. The E$_z$ vector fields are concentrated on the upper parts of the ring and grooves. The equivalent circuit of the spoof LSP has two parallel capacitances between each groove and is connected in series. In the equivalent circuit, when the capacitors are connected in parallel, the sensitivity of the loaded material is not affected by the loading position. However, in the case of series capacitance, the loading position is important for sensitivity. The series-connected capacitors have higher capacitance shifts when the sample is loaded only in the strong capacitance area. Thus, the sensitivity is higher in the capacitance-concentrated region than in the weak capacitance region. The area with the dotted lines refers to the strong capacitance concentration of the curved PDMS channel, designed specifically for increased sensitivity. Hence, only a 3.4 µL sample needs to be injected for measurement [28,29]. The reflected coefficients were used to assess the performance of the fabricated sensor. In the simulation, the resonant frequency and reflection coefficient of the designed sensor shifted in parallel to the changes in sample concentration. However, during the measurements, only the reflection coefficients were used because of the noise.

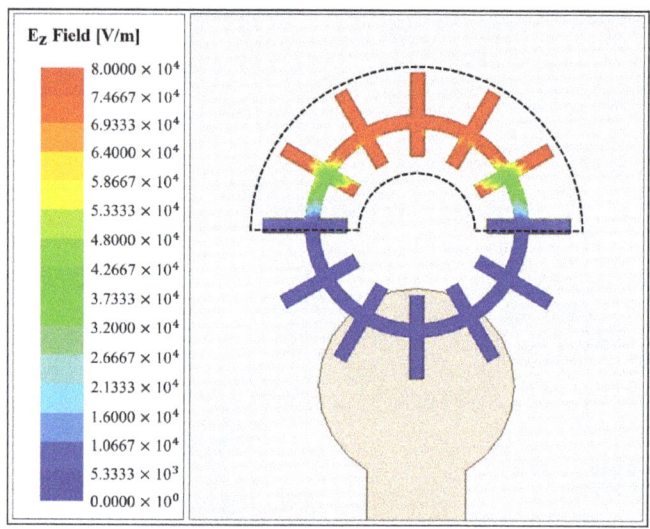

Figure 3. Simulated magnitudes of the E_z vectors of the E-field at a resonant frequency of 37.02 GHz.

3. Results and Discussion

3.1. Measurement Results

The fabricated glucose concentration sensor was evaluated using a vector network analyzer (VNA) with a 2.92 mm K-band connector. Figure 4 shows the simulated and measured results of the sensor. Figure 4a shows the simulated and measured values of the sensor without and with the bonding film. The measured value of the sensor has a resonant frequency of 37.02 GHz and a reflection coefficient of −33.38 dB. The calculated Q-factor of the measured sensor is 308.5. The measured results of the sensor with the bonding film have a 34.025 GHz resonant frequency and a −21.95 dB reflection coefficient. The calculated loaded Q-factor of the measured sensor with the bonding film is 87.3. The PDMS-channel designs used for the simulation and measurement with the bonding film are shown in Figure 2d. The simulated results of the sensor without the bonding film are similar to those with the bonding film. The simulated sensor has a resonant frequency of 34.035 GHz and a reflection coefficient of −31.47 dB. The calculated loaded Q-factor of the simulated sensor is 200.2. The simulated sensor with the bonding film has a 34.005 GHz resonant frequency and −27.07 dB reflection coefficient, with a Q-factor of 154.59. These differences in the resonant frequency can occur because of fabrication errors, such as in the groove widths and lengths. The Q-factor is calculated using Equation (3).

$$\frac{f_r}{f_{3dB\ (upper)} - f_{3dB\ (lower)}}. \qquad (3)$$

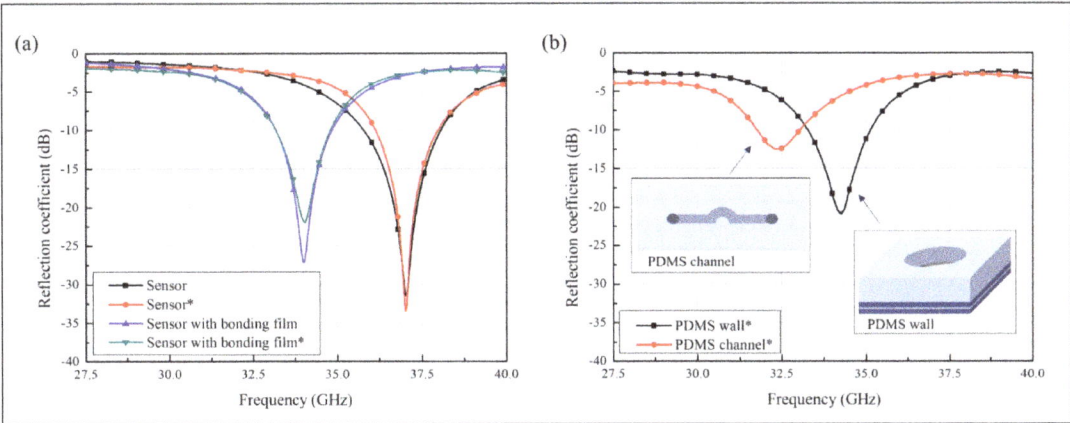

Figure 4. Simulated and measured results of (**a**) the sensor without or with the bonding films. (**b**) Measured results when the PDMS microfluidic channels were used (* is the measured result).

In Equation (3), f_r is the resonant frequency, and $f_{3dB\ (upper)}$ and $f_{3dB\ (lower)}$ are the higher and lower frequencies compared to the resonant frequency, which are 3 dB different from the resonant frequency. Thus, the denominator is the 3 dB bandwidth. The calculated Q-factor shows that the spoof LSP-resonator-based sensor has a high Q-factor, and the used ring pattern located at the center of the grooves has a higher Q-factor than that of the original spoof LSP resonator design with periodic grooves located at the interface of the circle [30].

Figure 4b shows the measured sensor results for two different microfluidic-channel shapes. The PDMS wall and channel are attached to the sensor through three-layer bonding films of the same size as each PDMS. The sensor with PDMS channels has a resonant frequency of 32.325 GHz and a reflection coefficient of −12.48 dB, and the sensor with the PDMS wall has a 34.255 GHz resonant frequency and a −20.9 dB reflection coefficient. The measured results show the tendencies of the resonant frequency and reflection coefficient shifts.

3.2. Sensitivity of the Proposed Glucose Sensor

The sensitivity of the sensor with the PDMS wall was evaluated using 1, 2, 5, 10, 50, and 100 mg/dL glucose solutions at 27.8 °C, which is the temperature at which the relative complex permittivities were measured. Likewise, the sensitivity of the sensor with the PDMS channel was evaluated using 1, 2, 5, 10, 50, 100, 200, and 300 mg/dL samples at 22.3 °C. Figure 5 shows the measured results of the sensor with the microfluidic channels filled with DI water or glucose-solution samples. In Figure 5a, the measured reflection coefficient differences depend on the injected samples (DI water and glucose solutions of 1, 2, 5, 50, and 100 mg/dL). The resonant frequency of the sensor with the PDMS wall filled with DI water was 24.75 GHz. For these measurements, an 8 μL sample solution was injected and extruded using a micropipette. The 1 mg/dL glucose solution yielded the same result as the 2 mg/dL sample. Thus, the sensor with the PDMS wall can distinguish a difference with a maximum of 2 mg/dL.

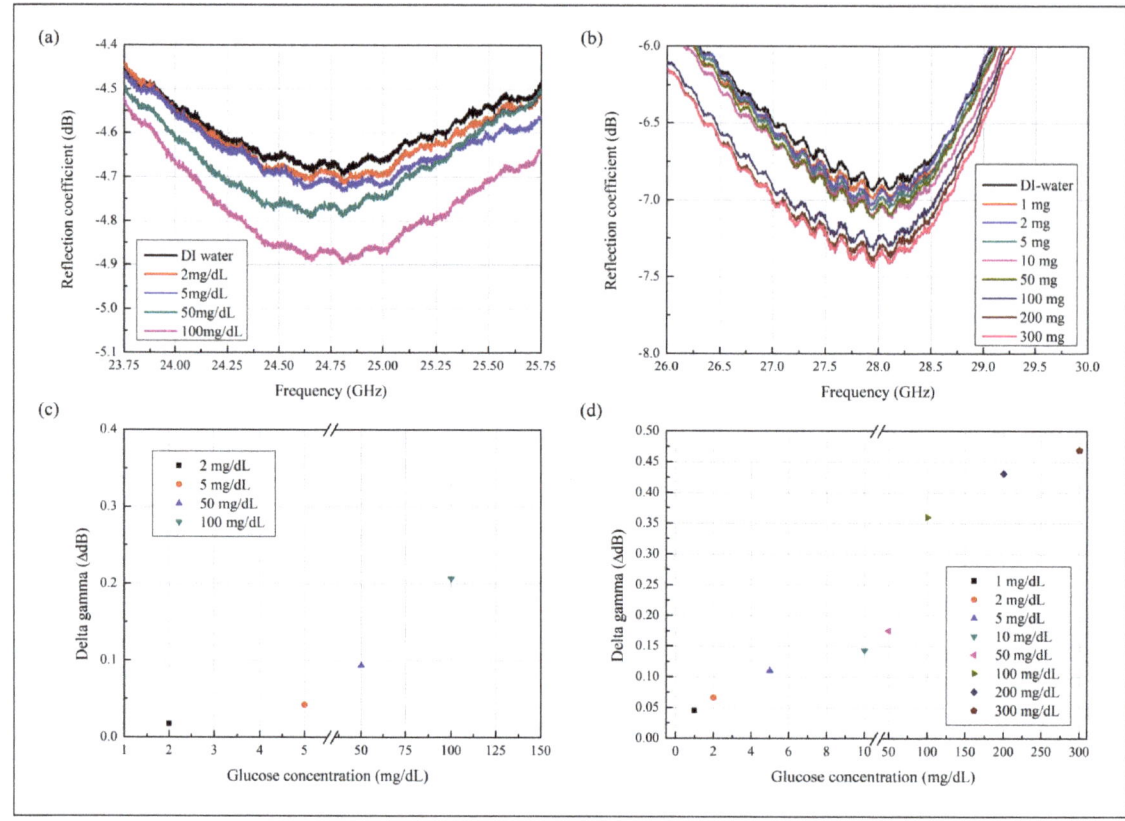

Figure 5. Measured reflection coefficients for the sensor with the (**a**) PDMS wall or (**b**) PDMS channel. Calculated delta gamma of the reflection coefficients for the sensor with the (**c**) PDMS wall at 24.8 GHz and 27.8 °C or (**d**) the PDMS channel at 28 GHz at 22.3 °C.

The reflection coefficient at 24.8 GHz was then used to confirm the sensing performance for glucose. The reflection coefficient change delta gamma has the largest value at 24.8 GHz. Figure 5b shows that the measured reflection coefficients of the sensor with the PDMS channel depend on the glucose concentration. The results shown correspond to DI water and 1, 2, 5, 10, 50, 100, 200, and 300 mg/dL glucose. The resonant frequency of the sensor with the PDMS channel filled with DI water was 28 GHz; hence, 28 GHz was used to assess the sensing performance with changes in glucose concentration. The graphs show that the sensor with the PDMS channel can distinguish a difference of 1 mg/dL in glucose concentration. Results from the PDMS wall and channel both show that when the glucose concentration increases, the amplitude of the reflection coefficient also increases.

Figure 5c,d shows the calculated sensing performances as the differences between the measured reflection coefficients. Figure 5c shows the delta gamma calculated using the reflection coefficients at 24.8 GHz. At 24.8 GHz, the sensor with the PMDS wall exhibited reflection coefficients from −4.68 to −4.89 with the glucose concentration increasing from 0 (DI water) to 100 mg/dL. More specifically, the measured reflection coefficients were −4.69, −4.72, −4.78, and −4.89 dB at 2, 5, 50, and 100 mg/dL, respectively. The delta gamma values were calculated according to the results obtained using DI water. The calculated values have differences of 0.01729, 0.045215, 0.09342, and 0.20637 dB at 2, 5, 50, and 100 mg/dL, respectively, based on the dB value of DI water. Thus, the sensor with the

PDMS wall has an average sensitivity of 525.17 dB/g/mL and a minimum detection limit of 2 mg/dL.

Figure 5d shows the assessment of the sensitivity by using the same method as in Figure 5c. The graph shows the calculated delta gamma values for the glucose concentrations. The results represent the calculated delta gamma values of the sensor with the PDMS channel at 28 GHz. The measured reflection coefficient of DI water was −6.89447 dB, and the values were −6.93943, −6.96078, −7.00517, −7.03714, −7.07025, −7.25414, −7.32565, and −7.36336 dB at 1, 2, 5, 10, 50, 100, 200, and 300 mg/dL, respectively. Thus, the calculated delta gamma values were 0.04496, 0.06631, 0.1107, 0.14267, 0.17578, 0.35967, 0.43118, and 0.46889 dB at 1, 2, 5, 10, 50, 100, 200, and 300 mg/dL according to the results obtained using DI water, respectively. The average sensitivity is 1566.9 dB/(g/mL), and the minimum detection limit is 1 mg/dL. These results show that the sensor can detect glucose concentrations by reflection coefficient shifts. The marked dotted lines shown in Figure 5c,d represent the linear ratio between delta gamma and glucose concentration. The inclination of the dot line shows the sensor with the PDMS channel has higher sensitivity than the sensor with the PDMS wall.

To investigate the sensor with either microfluidic channel, we used Equation (4).

$$Sensitivity = \frac{\Delta dB}{\Delta c} = \frac{|dB_{c1} - dB_{c2}|}{|c_1 - c_2|}. \tag{4}$$

The parameter c_n is glucose concentration (mg/dL), and dB_{cn} is the reflection coefficient of c_n at the resonant frequency of each result. The sensitivities are determined according to the reflection coefficients and sample concentration. The results show that the proposed sensor has an average sensitivity of 0.01567 dB/(mg/dL).

3.3. Analytical Characterization of the Sensor

To investigate the analytical characteristics of the sensor, we experimented and calculated the reusability, reproducibility, and response time. The PDMS channel is a highly reusable material. It is flexible and can be fabricated in various designs. In Figure 6a, the fabricated sensor has a different resonant frequency at the 60th iteration of removing the injected sample. Until the 60th trial, the measured results have the same resonant frequency and only have a different reflection coefficient. During the measurement, the PDMS channel, which is filled with air, can be affected by environmental factors, such as vibrations. Therefore, to measure the reusability, we investigated the changes in resonant frequency.

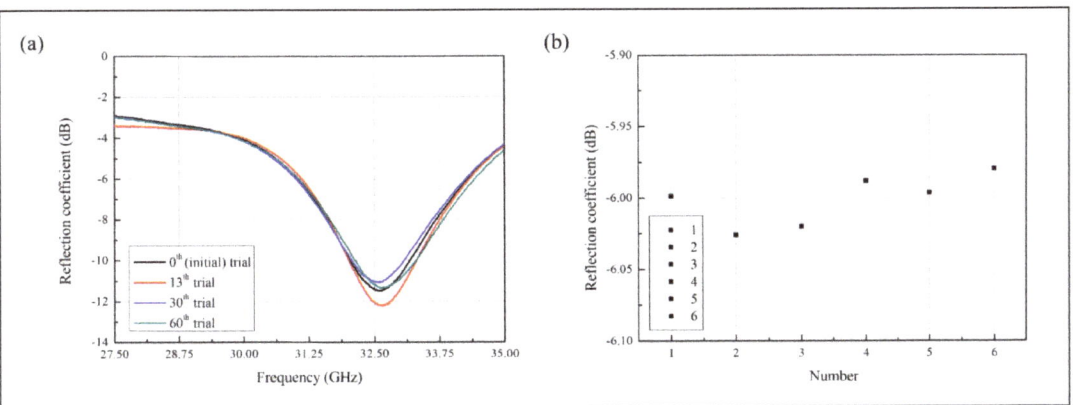

Figure 6. Measured reflection coefficients to investigate (**a**) the reusability and (**b**) the reproducibility of the sensor.

Figure 6b shows the reproducibility of the sensor. The 300 mg/dL concentration sample was used for measurement, and due to the difficulty of maintaining the temperature, it was measured at 25.2 °C. The average reflection coefficient was measured at −6.001431 dB, and the maximum and minimum values were −5.97966 and −6.02592 dB, respectively. The calculated average sensitivity was 0.01567 dB/(mg/dL), and the maximum and minimum values can be represented by 1.5 times the average 1 mg/dL sensitivity. Therefore, the sensor reproducibility relative standard deviation (RSD) is 0.3%. The RSD was calculated by the following Equation:

$$RSD = \frac{\text{standard deviration}}{\text{average}} \times 100\%. \quad (5)$$

During the measurement, we used the VNA and VNA-measured S-parameters in real-time. The conventional frequency-based glucose sensors, such as those based on the microwave or millimeter wave frequency, can monitor biological reactions in real-time. Therefore, real-time monitoring is one of the advantages of using the microwave sensing technology method [31].

3.4. Performance Comparison of the Sensors

To investigate the sensing performance, we compared the designed glucose sensors with the conventional sensors based on spoof SP or LSP, as well as other state-of-the-art glucose sensors, by using multiple parameters important for a sensor, such as physical size, sample volumes, minimum distinguishable concentration, and sensitivity. Table 1 shows the comparison with the sensors based on spoof SP or LSP, including those used to sense chemicals other than glucose. The designed glucose sensors are smaller, have higher Q-factors, and can measure smaller-sample volumes than the conventional spoof LSP sensors. The sizes of the designed millimeter-wave-based sensors are similar to those of the sensors based on the quarter-mode LSP. The fabricated sensor also has better sensitivity and a higher Q-factor compared to the quarter-mode spoof LSP sensor, which operates at microwave frequency. This observation indicates that the sensors based on millimeter-wave frequency show better performance than the LSP sensors. The former sensors are smaller, can assess smaller sample volumes, have higher sensitivity than the latter, and do not require additional circuit parts. Moreover, conventional spoof SP- or LSP-based sensors have higher sensitivity to permittivity than sensors based on other methods because of the near-field enhancement [32–35].

Table 1. Comparison with sensors using spoof SP or LSP.

Title [Ref. No]	Operating Frequency (GHz)	Physical Size (mm²)	Sample Volume (μL)	Q-Factor [1]	Distinguishable Concentration (mg/dL)	Sensitivity (dB/(g/mL))	Sensitivity (MHz/(g/mL))
This work	28	20 × 20	3.4	308.5	1	1566.9 [2]	N/A
[15]	3.16–3.76 [3]	20 [4] × 20 [4]	101.7	N/A	1164	N/A	773
[32]	6.86–7.8	17 × 17	3.9	25	N/A	940 MHz shift (10 to 90% ethanol)	
[33]	6.67	34 × 34	6 & 15.5	N/A	N/A	detection two chemicals	
[34]	1.5–2.5	42 × 40	12	40,000	9	N/A	29,111.1
[35]	8–12	52 × 24	N/A	N/A	25	N/A	200,000

[1] Measured value with only the sensor without any additional parts. [2] Calculated average value. [3] The value estimated from the plot. [4] Pattern size only.

In Table 2, the designed sensor is compared with other state-of-the-art glucose sensors. Other sensing techniques are also presented, such as coplanar waveguides (CPWs) with interdigital (IDT) structures and electric LC (ELCS) resonators, and complementary split-ring resonators (CSRRs). These sensors have miniaturized sizes despite operating at microwave frequencies. However, the required sample volume for sensing the glucose solution is different. For sensing the glucose concentration, these sensors need at least 5 times the sample volume needed in the proposed sensor. Thus, the proposed sensor has higher sensitivity and requires less sample volume than the other sensors [36–41]. However, microwave-based sensors have a limitation in selectivity because their detection is based on changes in the dielectric constant and loss tangent. For instance, when different mixed solutions have a similar dielectric constant and loss tangent, their frequency responses are similar so the target substance cannot be detected [42,43].

Table 2. Comparison with state-of-the-art glucose sensors.

Title [Ref. No]	Sensing Technique	Operating Frequency (GHz)	Physical Size (mm²)	Sample Volume (µL)	Q-Factor	Sensitivity (dB/(g/mL))	Sensitivity (MHz/(g/mL))
This work	Spoof LSP resonance	28	20 × 20	3.4	308.5	1566.9	N/A
[23]	CSRR	2.9	26 × 40	N/A	N/A	7.5	N/A
[36]	CPW with IDT	3.9–4.12 [2]	25.4 × 30	15	20 [1]	15.3	235.32
[37]	CPW with ELC	3.41	16 [3] × 16 [3]	20	N/A	3.73	N/A
[38]	CSRR driven by ISM radar	2.45	20 × 66	600	60 [1]	N/A	125,000
[39]	Hilbert curve	6	20.4 × 40.4	500	62	1560	N/A
[40]	Microstrip	5.5–6.7	80 × 80	14,000	81 [1]	N/A	54,000
[41]	WGM [4]	50–70	50 × 7.64 [3]	50–370	N/A	1000	N/A

[1] Estimated value from the plot. [2] dB parameter used frequency. [3] Only pattern size. [4] Whispering gallery modes.

Figure 7 shows the photos of the fabricated sensors and their measurement settings. Figure 7a shows the sensor with the top and bottom substrates attached using bonding films and connected to a 2.92 mm K-band connector. The white lines denote the position where the top substrate aligns with the PDMS wall. Figure 7b shows the top view of the sensor with the PDMS wall containing the glucose solution. Figure 7c shows the measurement settings using the VNA for the sensor with the PDMS channel.

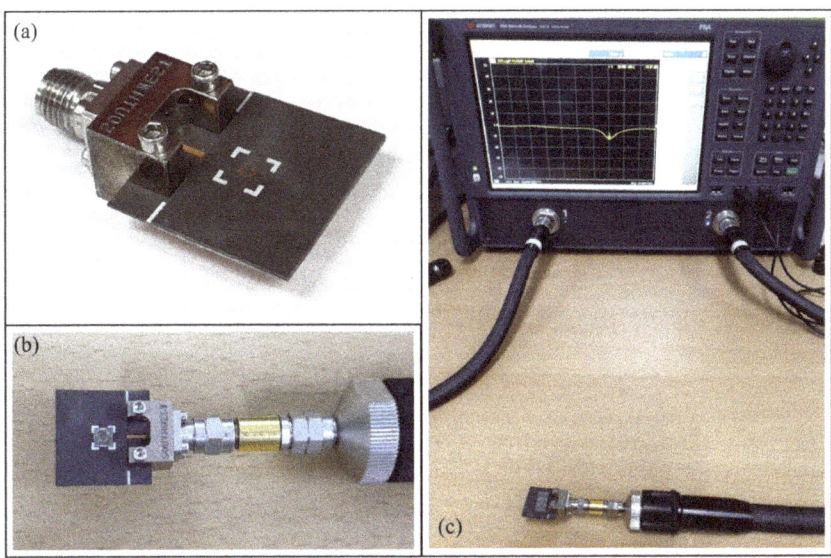

Figure 7. Photos of the fabricated sensor with the (**a**) connector and (**b**) PDMS wall filled with the glucose solution. (**c**) Measurement settings of the sensor with the PDMS channel.

4. Conclusions

In this study, we designed a spoof LSP resonator sensor that operates at millimeter-wave frequencies for sensing glucose concentrations. By measuring the complex relative permittivity of glucose solutions at various concentrations, we confirmed the dielectric properties of DI water and glucose solution. Glucose solutions in the range of 0–300 mg/dL were used for sensing, with a sample volume of 3.4 µL to investigate the sensor performance. The proposed sensor senses differences in glucose concentrations according to the changes in the reflection coefficients. The sensor using the PDMS channel with the multilayer bonding film has a sensitivity of 1566.9 dB/(g/mL), and it can distinguish a difference of 1 mg/dL. In addition, it can be reused 60 times and has a reproducibility of 0.3%. The proposed sensor was compared with other spoof LSP sensors and those using the microwave method. From these evaluations, it was observed that the proposed sensor has the advantages of being small, having high sensitivity, and the ability to work on small volumes of sample.

Author Contributions: Conceptualization, S.L., A.S. and Y.K.; methodology, A.S. and Y.K.; software, A.S. and Y.K.; validation, Y.K.; formal analysis, Y.K.; investigation, Y.K.; resources, Y.K.; data curation, Y.K.; writing—original draft preparation, Y.K.; writing—review and editing, S.L.; visualization, Y.K.; supervision, S.L.; project administration, S.L.; funding acquisition, S.L. All authors have read and agreed to the published version of the manuscript.

Funding: This work was supported by grants from the National Research Foundation of Korea (NRF), funded by the Korean government (MSIT) (2021R1A4A2001316 and 2020M3H4A3081832) and the Chung-Ang University Graduate Research Scholarship in 2020.

Institutional Review Board Statement: Not applicable.

Informed Consent Statement: Not applicable.

Data Availability Statement: The data presented in this study are available on request from the corresponding author.

Conflicts of Interest: The authors declare no conflict of interest.

References

1. Wang, P.Y.; Lu, M.S.C. CMOS thermal sensor arrays for enzymatic glucose detection. *IEEE Sens. J.* **2011**, *11*, 3469–3475. [CrossRef]
2. Cano, L.; Guti, J.; Mayoral, C.P.; Eduardo, L.P.; Pina, S.; Carrillo, L.T. Fiber optic sensors: A review for glucose measurement. *Biosensors* **2021**, *11*, 1–19.
3. Saptari, V.; Youcef-Toumi, K. Design of a mechanical-tunable filter spectrometer for noninvasive glucose measurement. *Appl. Opt.* **2004**, *43*, 2680–2688. [CrossRef]
4. Chretiennot, T.; Dubuc, D.; Grenier, K. Microwave-based microfluidic sensor for non-destructive and quantitative glucose monitoring in aqueous solution. *Sensors* **2016**, *16*, 1733. [CrossRef] [PubMed]
5. Madden, J.; Barrett, C.; Laffir, F.R.; Thompson, M.; Galvin, P.; O' Riordan, A. On-chip glucose detection based on glucose oxidase immobilized on a platinum-modified, gold microband electrode. *Biosensors* **2021**, *11*, 249. [CrossRef]
6. Gorst, A.; Zavyalova, K.; Mironchev, A. Non-invasive determination of glucose concentration using a near-field sensor. *Biosensors* **2021**, *11*, 62. [CrossRef] [PubMed]
7. Kumari, R.; Patel, P.N.; Yadav, R. An ENG resonator-based microwave sensor for the characterization of aqueous glucose. *J. Phys. D Appl. Phys.* **2018**, *51*. [CrossRef]
8. Govind, G.; Akhtar, M.J. Design of an ELC resonator-based reusable RF microfluidic sensor for blood glucose estimation. *Sci. Rep.* **2020**, *10*, 1–10. [CrossRef]
9. Li, Y.; Yao, Z.; Yue, W.; Zhang, C.; Gao, S.; Wang, C. Reusable, Non-invasive, and ultrafast radio frequency biosensor based on optimized integrated passive device fabrication process for quantitative detection of glucose levels. *Sensors* **2020**, *20*, 1565. [CrossRef]
10. Juska, V.B.; Pemble, M.E. A Critical review of electrochemical glucose sensing: Evolution of biosensor platforms based on advanced nanosystems. *Sensors* **2020**, *20*, 6013. [CrossRef]
11. Zhang, X.; Liu, Z. Superlenses to overcome the diffraction limit. *Nat. Mater.* **2008**, *7*, 435–441. [CrossRef] [PubMed]
12. Janković, N.; Ilić, S.; Bengin, V.; Birgermajer, S.; Radonić, V.; Alù, A. Acoustic spoof surface plasmon polaritons for filtering, isolation and sensing. *Results Phys.* **2021**, *28*, 104645. [CrossRef]
13. Atwater, H.A.; Polman, A. Erratum: Plasmonics for improved photovoltaic devices (Nature Materials (2010) 9 (205–213)). *Nat. Mater.* **2010**, *9*, 865. [CrossRef]

14. Khamsalee, P.; Mesawad, P.; Wongsan, R. Hybrid metamaterial for the secondary radar antenna system. *J. Electromagn. Eng. Sci.* **2020**, *20*, 221–233. [CrossRef]
15. Pandit, N.; Jaiswal, R.K.; Pathak, N.P. Plasmonic metamaterial-based label-free microfluidic microwave sensor for aqueous biological applications. *IEEE Sens. J.* **2020**, *20*, 10582–10590. [CrossRef]
16. Farokhipour, E.; Mehrabi, M.; Komjani, N.; Ding, C. A spoof surface plasmon polaritons (SSPPs) based dual-band-rejection filter with wide rejection bandwidth. *Sensors* **2020**, *20*, 7311. [CrossRef]
17. Zhang, Y.; Zhou, Y.J.; Cai, J.; Jiang, J.H. Amplification of spoof localized surface plasmons on active plasmonic metamaterials. *J. Phys. D Appl. Phys.* **2018**, *51*, 295304. [CrossRef]
18. Gao, Z.; Gao, F.; Xu, H.; Zhang, Y.; Zhang, B. Localized spoof surface plasmons in textured open metal surfaces. *Opt. Lett.* **2016**, *41*, 2181. [CrossRef]
19. Kianinejad, A.; Chen, Z.N.; Qiu, C.W. Design and modeling of spoof surface plasmon modes-based microwave slow-wave transmission line. *IEEE Trans. Microw. Theory Tech.* **2015**, *63*, 1817–1825. [CrossRef]
20. Zhang, X.; Cui, W.Y.; Lei, Y.; Zheng, X.; Zhang, J.; Cui, T.J. Spoof localized surface plasmons for sensing applications. *Adv. Mater. Technol.* **2021**, *6*, 1–24. [CrossRef]
21. Koutsoupidou, M.; Cano-Garcia, H.; Pricci, R.L.; Saha, S.C.; Palikaras, G.; Kallos, E.; Kosmas, P. Study and suppression of multipath signals in a non-invasive millimeter wave transmission glucose-sensing system. *IEEE J. Electromagn. RF Microw. Med. Biol.* **2020**, *4*, 187–193. [CrossRef]
22. Topfer, F.; Oberhammer, J. Millimeter-waves tissue diagnostics: The most promising fields for medical applications. *IEEE Microw. Mag.* **2015**, *16*, 97–113. [CrossRef]
23. Jang, C.; Park, J.K.; Lee, H.J.; Yun, G.H.; Yook, J.G. Temperature-corrected fluidic glucose sensor based on microwave resonator. *Sensors* **2018**, *18*, 3850. [CrossRef] [PubMed]
24. Hassan, R.S.; Park, S.I.; Arya, A.K.; Kim, S. Continuous characterization of permittivity over a wide bandwidth using a cavity resonator. *J. Electromagn. Eng. Sci.* **2020**, *20*, 39–44. [CrossRef]
25. Kim, K.C.; Kim, J.W.; Kwon, J.Y.; Kang, N.W. Characteristics of a cutoff cavity probe applicable to crack detection using the forced resonance microwave method. *J. Electromagn. Eng. Sci.* **2020**, *20*, 285–292. [CrossRef]
26. Andryieuski, A.; Kuznetsova, S.M.; Zhukovsky, S.V.; Kivshar, Y.S.; Lavrinenko, A.V. Water: Promising opportunities for tunable all-dielectric electromagnetic metamaterials. *Sci. Rep.* **2015**, *5*, 13535. [CrossRef]
27. Yang, B.J.; Zhou, Y.J.; Xiao, Q.X. Spoof localized surface plasmons in corrugated ring structures excited by microstrip line. *Opt. Express* **2015**, *23*, 21434. [CrossRef]
28. Yin, J.Y.; Ren, J.; Zhang, H.C.; Zhang, Q.; Cui, T.J. Capacitive-coupled series spoof surface plasmon polaritons. *Sci. Rep.* **2016**, *6*, 1–8. [CrossRef]
29. Memon, M.U.; Lim, S. Millimeter-wave chemical sensor using substrate-integrated-waveguide cavity. *Sensors* **2016**, *16*, 1829. [CrossRef]
30. Bao, D.; Rajab, K.Z.; Jiang, W.X.; Cheng, Q.; Liao, Z.; Cui, T.J. Experimental demonstration of compact spoof localized surface plasmons. *Opt. Lett.* **2016**, *41*, 5418. [CrossRef]
31. Mehrotra, P.; Chatterjee, B.; Sen, S. EM-Wave Biosensors: A Review of RF, microwave, mm-wave and optical sensing. *Sensors* **2019**, *19*, 1013. [CrossRef] [PubMed]
32. Shao, R.L.; Zhou, Y.J.; Yang, L. Quarter-mode spoof plasmonic resonator for a microfluidic chemical sensor. *Appl. Opt.* **2018**, *57*, 8472. [CrossRef]
33. Gholamian, M.; Shabanpour, J.; Cheldavi, A. Highly sensitive quarter-mode spoof localized plasmonic resonator for dual-detection rf microfluidic chemical sensor. *J. Phys. D Appl. Phys.* **2020**, *53*, 145401. [CrossRef]
34. Zhao, H.Z.; Zhou, Y.J.; Cai, J.; Li, Q.Y.; Li, Z.; Xiao, Z.Y. Ultra-high resolution sensing of glucose concentration based on amplified half-integer localized surface plasmons mode. *J. Phys. D Appl. Phys.* **2020**, *53*, 095305. [CrossRef]
35. Kandwal, A.; Nie, Z.; Igbe, T.; Li, J.; Liu, Y.; Liu, L.W.Y.; Hao, Y. Surface plasmonic feature microwave sensor with highly confined fields for aqueous-glucose and blood-glucose measurements. *IEEE Trans. Instrum. Meas.* **2021**, *70*, 1–9. [CrossRef]
36. Abedeen, Z.; Agarwal, P. Microwave sensing technique based label-free and real-time planar glucose analyzer fabricated on FR4. *Sens. Actuators A Phys.* **2018**, *279*, 132–139. [CrossRef]
37. Harnsoongnoen, S.; Wanthong, A. Coplanar waveguide transmission line loaded with electric-LC resonator for determination of glucose concentration sensing. *IEEE Sens. J.* **2017**, *17*, 1635–1640. [CrossRef]
38. Omer, A.E.; Shaker, G.; Safavi-Naeini, S.; Kokabi, H.; Alquié, G.; Deshours, F.; Shubair, R.M. Low-cost portable microwave sensor for non-invasive monitoring of blood glucose level: Novel design utilizing a four-cell CSRR hexagonal configuration. *Sci. Rep.* **2020**, *10*, 1–20. [CrossRef]
39. Odabashyan, L.; Babajanyan, A.; Baghdasaryan, Z.; Kim, S.; Kim, J.; Friedman, B.; Lee, J.H.; Lee, K. Real-time noninvasive measurement of glucose concentration using a modified hilbert shaped microwave sensor. *Sensors* **2019**, *19*, 5525. [CrossRef] [PubMed]
40. Sethi, W.T.; Issa, K.; Ashraf, M.A.; Alshebeili, S. In vitro analysis of a microwave sensor for noninvasive glucose monitoring. *Microw. Opt. Technol. Lett.* **2019**, *61*, 599–604. [CrossRef]
41. Omer, A.E.; Gigoyan, S.; Shaker, G.; Safavi-Naeini, S. WGM-based sensing of characterized glucose-aqueous solutions at mm-waves. *IEEE Access* **2020**, *8*, 38809–38825. [CrossRef]

42. Hosseini, N.; Baghelani, M. Selective Real-time non-contact multi-variable water-alcohol-sugar concentration analysis during fermentation process using microwave split-ring resonator based sensor. *Sens. Actuators A Phys.* **2021**, *325*, 112695. [CrossRef]
43. Characterization, M.D.; Ebrahimi, A.; Member, S.; Withayachumnankul, W. High-sensitivity metamaterial-inspired sensor for microfluidic dielectric characterization. *IEEE Sens. J.* **2014**, *14*, 1345–1351.

Article

Label-Free Detection and Spectrometrically Quantitative Analysis of the Cancer Biomarker CA125 Based on Lyotropic Chromonic Liquid Crystal

Hassanein Shaban [1,2], Mon-Juan Lee [3,4,*] and Wei Lee [1,*]

[1] Institute of Imaging and Biomedical Photonics, College of Photonics, National Yang Ming Chiao Tung University, Guiren District, Tainan 71150, Taiwan; hassanein.shaban@sci.asu.edu.eg
[2] Department of Basic Science, Faculty of Engineering, The British University in Egypt, El Sherouk City 11837, Egypt
[3] Department of Bioscience Technology, Chang Jung Christian University, Guiren District, Tainan 71101, Taiwan
[4] Department of Medical Science Industries, Chang Jung Christian University, Guiren District, Tainan 71101, Taiwan
* Correspondence: mjlee@mail.cjcu.edu.tw (M.-J.L.); wei.lee@nycu.edu.tw (W.L.)

Citation: Shaban, H.; Lee, M.-J.; Lee, W. Label-Free Detection and Spectrometrically Quantitative Analysis of the Cancer Biomarker CA125 Based on Lyotropic Chromonic Liquid Crystal. *Biosensors* **2021**, *11*, 271. https://doi.org/10.3390/bios11080271

Received: 16 July 2021
Accepted: 9 August 2021
Published: 11 August 2021

Publisher's Note: MDPI stays neutral with regard to jurisdictional claims in published maps and institutional affiliations.

Copyright: © 2021 by the authors. Licensee MDPI, Basel, Switzerland. This article is an open access article distributed under the terms and conditions of the Creative Commons Attribution (CC BY) license (https://creativecommons.org/licenses/by/4.0/).

Abstract: Compared with thermotropic liquid crystals (LCs), the biosensing potential of lyotropic chromonic liquid crystals (LCLCs), which are more biocompatible because of their hydrophilic nature, has scarcely been investigated. In this study, the nematic phase, a mesophase shared by both thermotropic LCs and LCLCs, of disodium cromoglycate (DSCG) was employed as the sensing mesogen in the LCLC-based biosensor. The biosensing platform was constructed so that the LCLC was homogeneously aligned by the planar anchoring strength of polyimide, but was disrupted in the presence of proteins such as bovine serum albumin (BSA) or the cancer biomarker CA125 captured by the anti-CA125 antibody, with the level of disturbance (and the optical signal thus produced) predominated by the amount of the analyte. The concentration- and wavelength-dependent optical response was analyzed by transmission spectrometry in the visible light spectrum with parallel or crossed polarizers. The concentration of CA125 can be quantified with spectrometrically derived parameters in a linear calibration curve. The limit of detection for both BSA and CA125 of the LCLC-based biosensor was superior or comparable to that of thermotropic LC-based biosensing techniques. Our results provide, to the best of our knowledge, the first evidence that LCLCs can be applied in spectrometrically quantitative biosensing.

Keywords: lyotropic chromonic liquid crystal; label-free biosensor; optical biosensor; protein assay; immunoassay; transmission spectrometry

1. Introduction

Most liquid crystal (LC)-based biosensing techniques reported to date employ thermotropic LCs, especially the rod-like nematic 4-cyano-4′-pentylbiphenyl (5CB), as the predominant sensing medium [1–4]. Building on technologies of 5CB biodetection at the LC–water interface, in the form of LC film on water, LC droplets, and LC emulsions [5,6], or the LC–glass interface with an LC cell configuration [7], various biosensors utilizing other types and phases of LCs, such as cholesteric LCs [8], dye LCs [9], and dual-frequency LCs [10], were developed to transduce the optical, electro-optical, and dielectric signals produced by biomolecules or biomolecular interactions [11]. Nevertheless, the biosensing application of lyotropic LCs is relatively scarce due possibly to the concentration-dependent polymorphic phase transitions and the lack of effective signal transduction approaches. Lyotropic chromonic liquid crystals (LCLCs) are a unique class of lyotropic LCs most frequently investigated in biosensing. LCLCs consist of water-soluble compounds characterized by a plank-like structure with a polyaromatic core linked to plural hydrophilic side groups (Figure 1a) [12]. When the drug disodium cromoglycate (DSCG) is dissolved in water, the nematic LCLC phase can be observed at room temperature (23 °C) at concentrations ranging from 12 to 17 wt%, with the viscosity strongly dependent on the DSCG content [13]. The DSCG molecules form rod-like columnar assemblages through π–π stacking of the hydrophobic polyaromatic structure, which are separated by a distance of ∼3.4 Å by virtue of the ionic repulsion between hydrophilic side groups [14]. Depending on the concentration and temperature, the self-assembled columnar stack of DSCG exhibits a lateral separation of ∼35 Å to 42 Å and lengthens with increasing DSCG concentration, contributing to a higher birefringence and lower viscosity than other lyotropic liquid crystals [14–16].

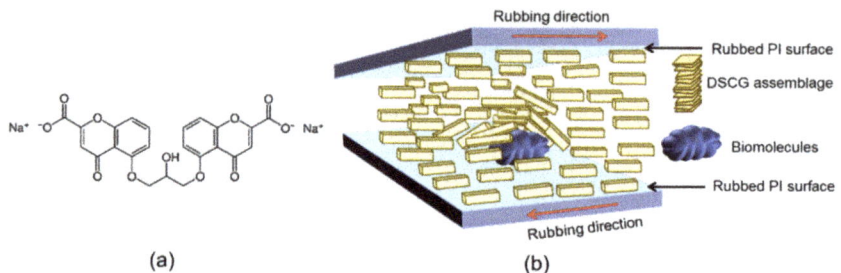

Figure 1. LCLC-based biosensing with planarly aligned DSCG LCLC. (**a**) The chemical structure of DSCG and (**b**) an illustration simulating the disturbance in the planarly aligned columnar stacks of DSCG in the presence of biomolecules.

Owing to its low cytotoxicity and hydrophilic nature, the self-organization and anisotropic properties of the nematic phase of DSCG were utilized in the dynamic detection of motile bacteria [17–20]. DSCG is more biocompatible with viruses and mammalian cells, allowing the vesicular stomatitis virus to remain active and replicate in human cervical epithelial carcinoma cells, in contrast to the zwitterionic surfactant C14AO, an amphiphilic lyotropic liquid crystal that results in virus inactivation and cell death [21]. In biomolecular detection, DSCG is considered congruent with molecular interactions without affecting the binding activity of anti-immunoglobulin (IgG) antibodies to immobilized human IgG [15]. The intensity of transmitted light observed under a polarizing optical microscope (POM) was enhanced when the ordered planar alignment of nematic DSCG was disrupted by the aggregation of streptavidin-coated latex beads in the presence of anti-streptavidin antibodies, but not in the absence of immunocomplex formation [22]. This implies that biodetection with the nematic phase of LCLCs is governed by a similar principle to thermotropic LCs, in which the level of disturbance caused by the analyte transduces to the amplitude of the resulting optical signal (Figure 1b).

The cancer antigen 125 (CA125) is expressed as a membrane glycoprotein on the cell surface of ovarian, breast or gastric cancer cells, but may be released in soluble form into the blood. The reference level of CA125 in a blood sample is 30–35 U/mL, beyond which indicates a higher risk of cancer progression [23]. Conventional methods for the detection of CA125 are sandwich assays, a type of immunoassay in which CA125 reacts specifically with a capture antibody immobilized on a solid substrate and a detection antibody with fluorometric or colorimetric labeling. The procedure of such label-based detection is time-consuming and the nonspecific binding of the labeled antibody may lead to false-positive signals. As part of an effort to develop rapid screening and cost-effective point-of-care diagnostics, novel techniques for the label-free detection of CA125 have been actively investigated, including electrochemical immunosensors [24,25], aptasensors based on field-effect transistors [26], and impedimetric immunosensors with gold nanostructured screen-printed electrodes [27]. In biosensing on the basis of thermotropic LCs, our previous work demonstrated the feasibility of a nematic LC of high birefringence as well as an LC–

photopolymer composite in highly sensitive label-free CA125 immunodetection [28–30]. Because current LC-based biosensing technologies rely heavily on the optical response derived from the LC texture observed under a POM, which can only provide qualitative or semiquantitative results, we incorporated transmission spectrometric analysis in a label-free CA125 immunoassay employing dye-doped LC as the sensing medium to enhance detection sensitivity and to elaborate a more accurate quantitative strategy [31].

In this study, an LCLC-based quantitative protein biosensor and an immunosensor utilizing the nematic phase of DSCG as the sensing medium were developed in conjunction with transmission spectrometry. The sensing platform was fabricated by sandwiching the aqueous solution of DSCG between a pair of glass substrates coated with polyimide (PI) to promote planar alignment. The optical signal derived from the biological analyte—either the common protein standard BSA or the tumor marker CA125 captured by the anti-CA125 antibody—immobilized at the LC-glass interface was analyzed by measuring the transmission spectra of the sandwich cell placed between parallel or crossed polarizers. The spectrometric analysis thus established provides not only absolute quantitation of analyte concentration from the wavelength-dependent optical response, but also the basis for the optimization of detection sensitivity and limit of detection.

2. Materials and Methods

2.1. Materials

Optical-grade glass substrates (22 × 18 × 1.1 mm) were purchased from Ruilong Glass, Miaoli, Taiwan. The commercial PI SE-150 (0821) from Nissan Chemical was utilized to prepare a planar alignment layer on a glass substrate. Both DSCG and BSA were provided by Sigma-Aldrich. At 14 wt% in DI water, DSCG exhibits the nematic phase, and the phase transition temperature from the nematic to isotropic phase was determined to be ca. 27.5 °C. Recombinant human CA125/MUC16 protein and anti-CA125 antibody employed in the LCLC-based immunoassay were manufactured by R&D Systems and Santa Cruz Biotechnology, respectively.

2.2. Preparation of PI-Coated Glass Substrates

Cleaned glass substrates were spin-coated with PI SE-150 (0821) at 2000 rpm for 30 s, followed by 6000 rpm for 60 s. The PI-coated substrates were then soft-baked at 85 °C for 10 min and successively hard-baked at 240 °C for 60 min. Unidirectional rubbing was performed 30 times on the PI-coated surface to enhance the planner anchoring strength, as shown in Figure 2a,b.

2.3. Construction of the LCLC-Based Protein Detection Platform

The surface of the rubbed PI-coated substrates was dissected into two areas: one for immobilizing biomolecules and detecting the analyte, and the other for monitoring background signals in the absence of immobilized proteins. BSA or anti-CA125 antibody was dispensed at 30 µL/spot with a Gilson Pipetman G P200G micropipette on one of the PI-coated glass substrates, which was dried at 30 °C for 30 min and rinsed with DI water to remove non-adsorbed biomolecules (Figure 2c,d). The circular area of the immobilized protein thus formed covered the entire cross section of the propagating light beam during the optical measurement. To assemble the LC cell, rod spacers of 15 µm in diameter were mixed with a small amount of the epoxy resin AB glue and distributed on two parallel sides of the glass substrate with immobilized biomolecules, which was covered with another protein-free PI-coated substrate with antiparallel rubbing direction (Figure 2e,f). The assembled LC cell was allowed to dry for 5 min at room temperature before further experiments were carried out. The cell gap of the empty LC cell was determined by optical interferometry with the Ocean Optics HR2000+ high-resolution USB fiber-optic spectrometer [32]. Afterwards, the LC cell was filled with 10-µL aqueous solution of 14-wt% DSCG by suction and sealed with AB glue (Figure 2g).

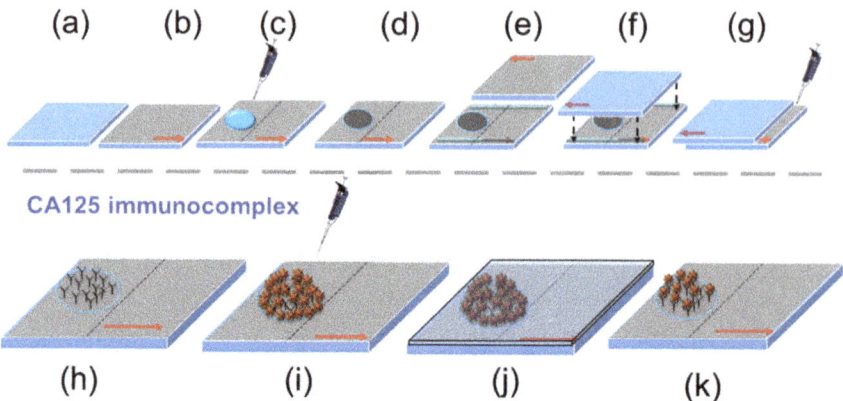

Figure 2. The LCLC-based protein detection and immunoassay platform. A cleaned optical glass substrate (**a**) was coated with PI SE-150 and rubbed unidirectionally (**b**). A 30 µL aqueous solution of BSA or anti-CA125 antibody was dispensed on the rubbed PI-coated surface (**c**) and allowed to dry at 30 °C for 30 min (**d**). After rinsing with DI water, the 15-µm spacer was mixed with the epoxy resin AB glue and distributed on two parallel edges of the optical glass substrate (**e**), followed by LC cell assembly with another protein-free PI-coated glass substrate with antiparallel rubbing direction (**f**). Finally, the LC cell was filled with 14-wt% DSCG solution for further optical measurements (**g**). For the CA125 immunoassay, immobilized anti-CA125 antibodies (**h**) were reacted with CA125 by dispensing 30 µL of the CA125 protein solution on the glass substrate (**i**), which was covered with a cover glass to allow immunoreaction to occur for 30 min (**j**). After removing the cover glass, the glass substrate was rinsed with DI water (**k**), followed by LC cell assembly as described in (**e**–**g**).

2.4. LCLC-Based ca125 Immunoassay

For the CA125 immunoassay, the immobilized anti-CA125 antibody was further reacted with the CA125 protein by dispensing 30 µL of the CA125 solution on the glass substrate, followed by covering the glass substrate with a cover glass so that the entire glass surface was in contact with the CA125 protein solution (Figure 2h–j). After reacting for 30 min, the cover glass was removed, and the glass substrate was rinsed with DI water to eliminate unbound CA125 (Figure 2k). LC cell assembly was then performed as described in Section 2.3 as well as illustrated in Figure 2e–g.

2.5. Transmission Spectrometric Measurements

Transmission spectra in the wavelength range of 400–800 nm were acquired with an Ocean Optics HR2000+ high-resolution fiber-optic spectrometer equipped with an Ocean Optics HL-2000 tungsten halogen as the light source. During spectral measurements, the LC cell was placed between two linear polarizers, with its rubbing direction parallel to the transmission axis of the lower polarizer or the analyzer, as shown in Figure 3. The transmission axis of the upper polarizer was oriented by either 0° or 90° measured from that of the lower polarizer, forming two modes of spectrometric detection with parallel or crossed polarizers, respectively.

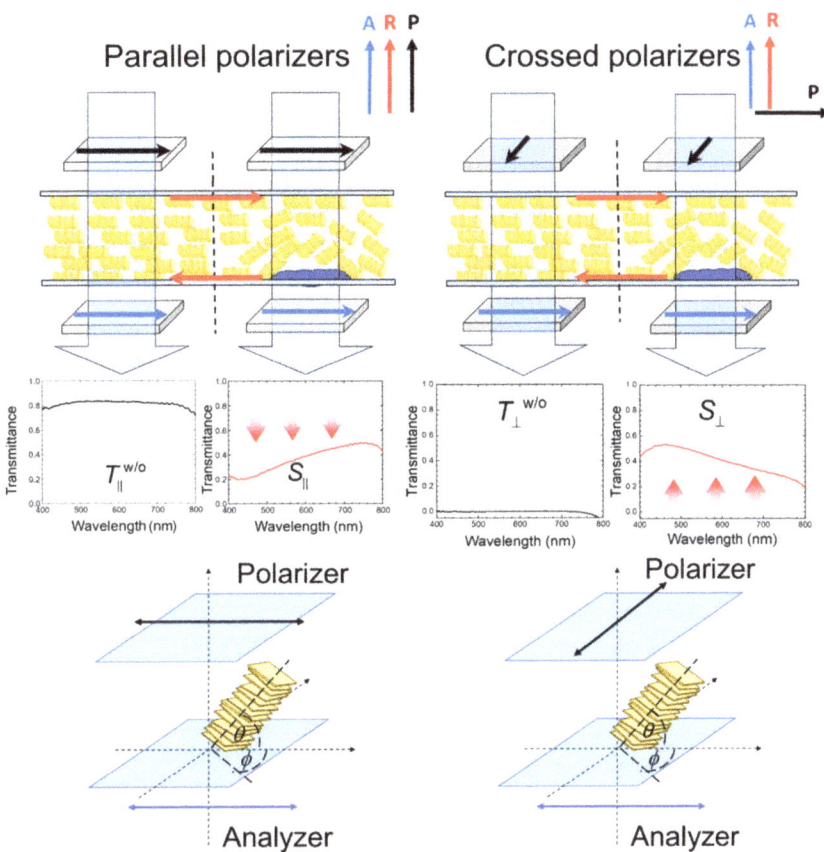

Figure 3. Transmission measurements with parallel and crossed polarizers in the LCLC-based biosensor. When biomolecules are immobilized at the LC–glass interface, leading to disturbance in the planar alignment of LCLC, transmittance decreases when analyzed with parallel polarizers (left panel) and increases when analyzed with crossed polarizers (right panel).

2.6. Optical Texture Observation

BSA immobilization and formation of the CA125 immunocomplex were enabled as described in Sections 2.3 and 2.4 except that BSA or anti-CA125 antibody was immobilized at 3 µL/spot to form a 3 × 3 array on the PI-coated glass substrate. The optical texture of the DSCG aqueous solution was observed under a POM (Olympus BX51, Tokyo, Japan) with the rubbing direction of the LC cell parallel to one of the transmission axes of the crossed polarizers.

3. Results and Discussion

3.1. Principle of Detection by Transmission Spectrometry in the LCLC-Based Biosensor

When transmission spectrometric analysis is performed with parallel polarizers, the transmittance or normalized intensity of the transmitted light, $I_\parallel$, can be formulated as:

$$I_\parallel = 1 - \sin^2(2\phi) \cdot \sin^2(\delta/2) \tag{1}$$

where ϕ is the azimuthal angle of the average molecular axis of LC, or the LC director, with respect to the transmission axis of the analyzer, and δ is the phase retardation [13]. The phase retardation can be, in turn, calculated as $\delta = 2\pi d \Delta n / \lambda$, where d is the cell gap, λ is the wavelength of the incident light, and Δn is the LC birefringence. By definition, $\Delta n \equiv n_{\text{eff}} - n_\perp$, where $n_\perp$ is the refractive index of the LC when the electric field of the incident light is perpendicular to the LC director, while n_{eff} is the effective refractive index given by:

$$n_{\text{eff}} = \frac{n_\perp n_\parallel}{\sqrt{n_\parallel^2 \sin^2 \theta + n_\perp^2 \cos^2 \theta}},\qquad(2)$$

in which $n_\parallel$ is the refractive index of the LC when the vibration direction of the electric field component of the impinging light is parallel to the LC director, and θ is the pretilt angle between the director and the substrate plane. In the absence of an analyte, the LC director is parallel to the transmission axes of the polarizers ($\phi = 0°$) in the parallel polarizer scheme so that $I_\parallel$ is at its maximum ($I_\parallel = 1$). When the orientation of LCs is disturbed by biomolecules immobilized at the LC–glass interface, ϕ grows ($\phi > 0°$), resulting in a decrease in $I_\parallel$. As $\phi = 45°$, Equation (1) becomes $I_\parallel = \cos^2(\delta/2)$, and no optical transmission is observed at $d\Delta n / \lambda = m + 1/2$ (where $m = 0$ or a positive integer).

Conversely, when spectral measurements are performed with crossed polarizers in the absence of an analyte, the transmittance or normalized intensity of the transmitted light, $I_\perp$, can be expressed as:

$$I_\perp = \sin^2(2\phi) \cdot \sin^2(\delta/2),\qquad(3)$$

which is therefore at its minimum when $\phi = 0°$. In the presence of an analyte when $\phi > 0°$, $I_\perp$ increases and maximum optical transmission results with $\phi = 45°$ such that Equation (3) becomes $I_\perp = \sin^2(\delta/2)$ at $d\Delta n / \lambda = m + 1/2$ (where $m = 0$ or a positive integer). Accordingly, when biomolecules are present on the lower PI-coated glass substrate of the LC cell, the planar alignment of LC molecules in contact with and in close proximity to the analyte is disturbed, and the wavelength-dependent optical signal obtained with either parallel or crossed polarizers is generated at non-zero azimuthal angles ($\phi \neq 0$) and reaches its maximum at $\phi = 45°$ (Figures 1 and 3). Moving toward the upper PI-coated substrate of the LC cell, where no biomolecules are immobilized, the homogeneous alignment of LC molecules is preserved. The LC molecules in the region with immobilized biomolecules are therefore in the twisted configuration, surrounded by those in perfectly planar state in areas without the analyte.

Moreover, the value of θ varies between $\theta = 0°$, corresponding to the planar alignment of the LC at maximum birefringence, and $\theta = 90°$, signifying the vertical alignment of the LC with vanished birefringence ($\Delta n = 0$). At $\theta = 0°$, the deviation of ϕ from zero ($\phi > 0$) strongly affects $I_\parallel$ and $I_\perp$, which, in contrast, remain unaltered by ϕ at $\theta = 90°$. According to the molecular theory of surface tension, the surface tension of the solid substrate (γ_c) is much greater than the surface tension of the planarly aligned LC (γ_l). Therefore, in the presence of an analyte when $\phi > 0$, the LCLC may be directed by surface tension to become reoriented outside the unidirectional "grooves" produced by rubbing. In addition, the twist elastic constant K_{22} of LCLC is much smaller than the splay and bend elastic constants, K_{11} and K_{33}, respectively, supporting that the orientation of LCLC molecules tends to deviate azimuthally in the substrate plane in response to external stimuli [33].

To quantitate the optical signal in correlation with the amount of analyte and to eliminate false-positive or nonspecific background noise, the reduced transmittance parameters obtained with parallel and crossed polarizers, $T_\parallel$ and $T_\perp$, are defined as:

$$T_{ll} = \frac{S_{ll} - T_\perp^{w/o}}{T_{ll}^{w/o} - T_\perp^{w/o}}\qquad(4)$$

and

$$T_\perp = \frac{S_\perp - T_\perp^{w/o}}{T_\parallel^{w/o} - T_\perp^{w/o}} \qquad (5)$$

respectively, where $S_\parallel$ and $S_\perp$ stand for the respective transmittance in the presence of analyte molecules measured with parallel and crossed polarizers, and $T_\parallel^{w/o}$ and $T_\perp^{w/o}$ are the transmittance in the absence of an analyte measured with parallel and crossed polarizers, respectively.

3.2. LCLC-Based Spectrometric Quantitation of BSA

In this work, the protein detection capability of the LCLC-based biosensor was demonstrated with a globular protein, the common protein standard BSA, and an antibody, the anti-CA125 antibody against the cancer biomarker CA125. Various concentrations of BSA were first immobilized on the PI-coated glass substrate, followed by LC cell assembly and optical texture observation under a POM with crossed polarizers. In the absence of BSA, DSCG was planarly aligned by the planar alignment agent PI, and the optical texture was completely dark, as depicted in Figure 4. As BSA accumulated at the LC-glass interface, light leakage increased with increasing concentrations of BSA (Figure 4). A similar dark-to-bright transition was also reported in the detection of BSA by homeotropically aligned DSCG [34]. However, DSCG tends to reorient from the homeotropic to the planar state over time [35]. Indeed, because of the low anchoring energy of conventional alignment agents and rubbing methods, it is more difficult to align hydrophilic LCLCs than hydrophobic thermotropic LCs, especially for the homeotropic alignment of LCLCs [36,37]. Our method of planar alignment with rubbed PI expectedly offered stable alignment with promoted anchoring strength and uniform surface [38].

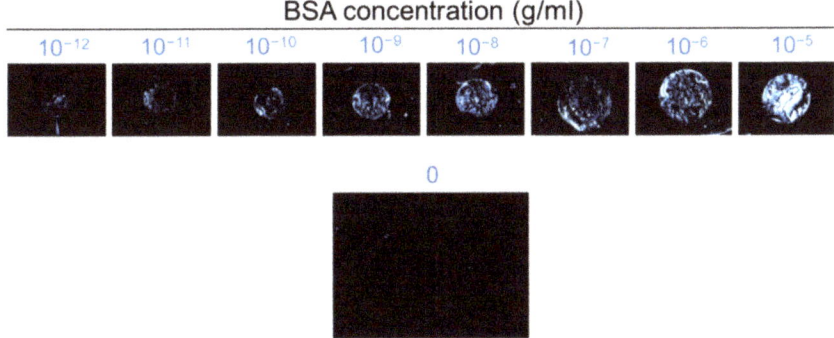

Figure 4. Optical textures of planarly aligned LCLC in the presence of BSA. The LCLC-based protein detection was performed with an aqueous solution of 14-wt% DSCG as the sensing medium in contact with 10^{-12}–10^{-5} g/mL BSA immobilized on the PI-coated glass surface. The micrograph at the bottom displays a completely dark texture of a reference of dried DI water containing no BSA.

Transmission spectrometric measurement with parallel and crossed polarizers was performed to explore methods for the absolute quantitation of the optical signal (Figure 5). As shown in Figure 5a, when parallel polarizers were utilized, reduced transmittance in the wavelength range of 400–800 nm decreased with increasing concentrations of BSA. At 10^{-12} g/mL, $T_\parallel$ was close to its maximum value of unity, which corresponds to the planar state in the absence of an analyte, suggesting that the amount of BSA was insufficient to significantly alter the orientation of DSCG assemblages in water. As the concentration of BSA increased, $T_\parallel$ decreased in a concentration-dependent manner, indicating an increase in the azimuthal angle and the deviation of the DSCG alignment from the rubbing direction. On the other hand, when crossed polarizers were applied, $T_\perp$ at 10^{-12} g/mL BSA approached its minimum of 0, which represents unperturbed planar alignment, and the increase in the amount of BSA led to elevated $T_\perp$ in a concentration-dependent manner (Figure 5b).

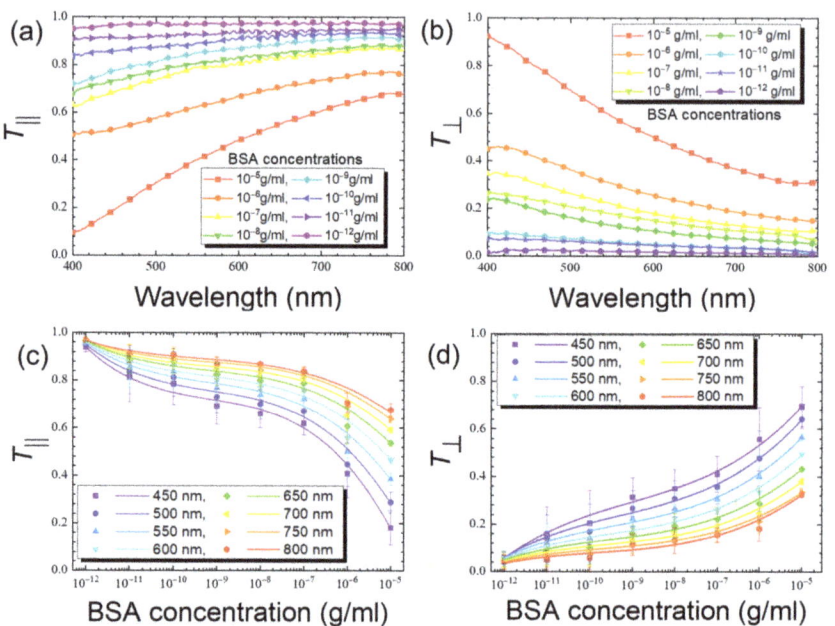

Figure 5. Reduced transmission data of LCLC with 14-wt% DSCG in the presence of BSA at various concentrations ranging from 10^{-12} to 10^{-5} g/mL. The reduced transmittance parameters (**a**) $T_\parallel$ and (**b**) $T_\perp$ in the wavelength range of 400–800 nm at various BSA concentrations and (**c**) $T_\parallel$ and (**d**) $T_\perp$ plotted against BSA concentration at selected wavelengths of 450, 500, 550, 600, 650, 700, 750, and 800 nm for quantitative purpose. $T_\parallel$ and $T_\perp$ correspond to the respective parallel polarizer and crossed polarizer schemes for the spectrometric detection, respectively. Error bars represent standard deviations ($n \geq 3$).

When the reduced transmittance $T_\parallel$ ($T_\perp$) at selected wavelengths was plotted against BSA concentration, a negative (positive) correlation was found (Figure 5c,d). While linear correlation can be ascertained in a narrower range of concentrations, the calibration curve for the entire BSA concentration range (10^{-12}–10^{-5} g/mL) was obtained through the third-order polynomial curve fitting that describes $T_\parallel$ or $T_\perp$ as a function of the logarithm of BSA concentration. The quality of the curve fitting, as evaluated by calculating the coefficient of determination, R^2, is satisfactory at ~0.99 at all wavelengths examined. One can see from (Figure 5c,d) that a more significant decrease in $T_\parallel$ and increase in $T_\perp$ with increasing BSA concentrations, respectively, were observed for a shorter wavelength, especially at 450 nm. These results indicate that the sensitivity of the LCLC-based spectrometric protein quantitation can be fine-tuned through the wavelength of the incident light, and the highest sensitivity was found at the shortest wavelength of 450 nm examined in this study. This can be explained by the effective phase retardation δ, which was larger at a shorter wavelength, in accordance with the wavelength-dependent birefringence of DSCG [13]. The limit of detection (LOD) was calculated using the following equation:

$$\text{LOD} = \frac{3s}{m} \qquad (6)$$

where s is the standard deviation of the y-intercept and m is the slope of the calibration curve obtained by linear regression [39]. Consequently, the LOD obtained with parallel polarizers ($LOD_{\parallel}$) and that with crossed polarizers ($LOD_{\perp}$) at all measured wavelengths ranged between 10^{-11} and 10^{-10} g/mL BSA, with $LOD_{\parallel}$ and $LOD_{\perp}$ values at 450 nm calculated as 2.4×10^{-10} and 6.0×10^{-11} g/mL BSA, respectively.

3.3. LCLC-Based Spectrometric Quantitation of the Anti-CA125 Antibody

The effect of the concentration of the anti-CA125 antibody on the optical texture of DSCG is shown in Figure 6. Compared with the optical texture of BSA as presented in Figure 4, the brightness of the POM image for the anti-CA125 antibody was less intense, presumably due to the higher molecular weight of the anti-CA125 antibody (150 kDa) with respect to BSA (66.5 kDa), and thus a smaller amount of the anti-CA125 antibody was present at the LC–glass interface at the same mass concentration as BSA to induce the reorientation of DSCG. The difference in the level of disturbance may also be related to the interaction between DSCG and the analyte in an aqueous environment. It was reported that DSCG may bind to BSA through electrostatic interaction and alter the conformation of the protein [40], but how DSCG interacts with the anti-CA125 antibody remains to be investigated.

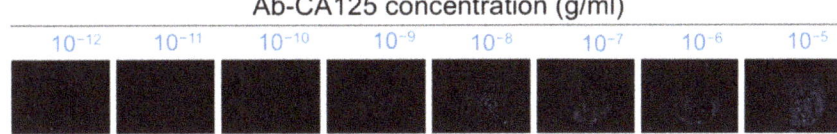

Figure 6. Optical textures of planarly aligned LCLC in the presence of the anti-CA125 antibody. The LCLC-based protein detection was performed with an aqueous solution of 14-wt% DSCG as the sensing medium in contact with 10^{-12}–10^{-5} g/mL anti-CA125 antibody immobilized on the PI-coated glass surface.

The transmission spectra obtained with parallel or crossed polarizers at various concentrations of the anti-CA125 antibody, as well as the results of the third-order polynomial regression of the calibration curve for the antibody, with R^2 values ranging from 0.94 to 0.99, are illustrated in Figure 7. As seen in Figure 5 for BSA, the trends of the concentration-dependent decrease in $T_{\parallel}$ and increase in $T_{\perp}$ were also detected for the anti-CA125 antibody (Figure 7). However, the change in either reduced transmittance parameter with the concentration of anti-CA125 antibody was less pronounced in comparison with that for BSA, suggesting that the level of disturbance and the increase in ϕ caused by the antibody were less significant, in agreement with our observation in the optical texture (Figure 6). The $LOD_{\parallel}$ calculated from Equation (6) for the anti-CA125 antibody based on the calibration curve in Figure 7c was higher than the $LOD_{\perp}$ derived from Figure 7d at all selected wavelengths, with $LOD_{\parallel} = 1.2 \times 10^{-10}$ g/mL and $LOD_{\perp} = 6.2 \times 10^{-12}$ g/mL as determined at 450 nm.

To further explore the anti-CA125 antibody data, the $T_{\parallel}/T_{\perp}$ values at various wavelengths were calculated and plotted against the concentration of the anti-CA125 antibody and a linear correlation resulted (Figure 8). The absolute value of $T_{\parallel}/T_{\perp}$ was significantly larger than $T_{\parallel}$ or $T_{\perp}$ alone, decreasing with increasing analyte concentration, and was higher at longer wavelengths. Transforming the nonlinear behavior in Figure 7c to a linear correlation in Figure 8 thus amplifies the optical signal and its variation with analyte concentration through data processing, especially at longer wavelengths, where the detection sensitivity was inferior when only $T_{\parallel}$ was considered. This allows quantitative analysis to be performed at less optimized conditions such as longer wavelengths or in the presence of a small amount of analytes. The LOD calculated from Equation (6) for the anti-CA125 antibody based on the calibration curve at 800 nm in Figure 8 was 2.6×10^{-11} g/mL.

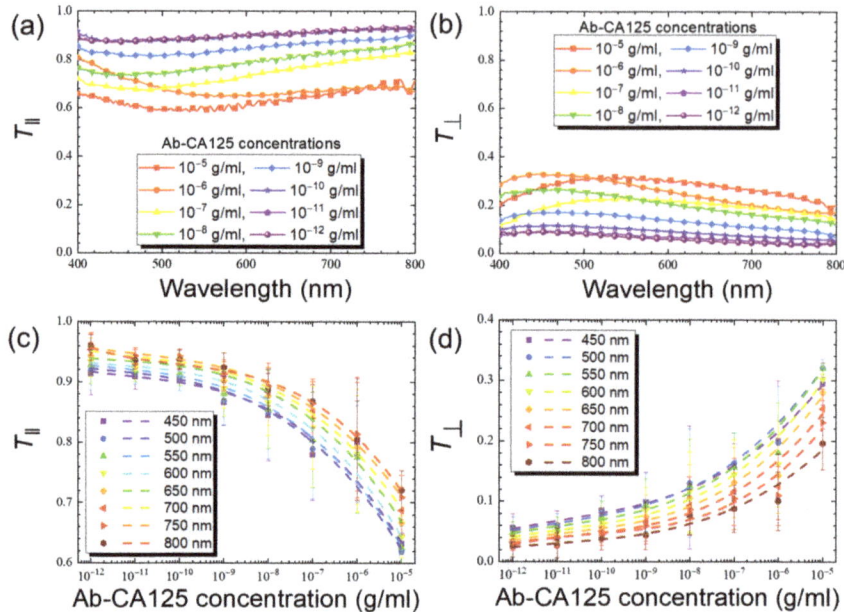

Figure 7. Reduced transmission data of LCLC with 14-wt% DSCG in the presence of the anti-CA125 antibody at various concentrations ranging from 10^{-12} to 10^{-5} g/mL. The reduced transmittance parameters (**a**) $T_\parallel$ and (**b**) $T_\perp$ in the wavelength range of 400–800 nm at various antibody concentrations and (**c**) $T_\parallel$ and (**d**) $T_\perp$ plotted against antibody concentration at selected wavelengths of 450, 500, 550, 600, 650, 700, 750, and 800 nm for quantitative purpose. $T_\parallel$ ($T_\perp$) corresponds to the measurements with parallel polarizers (crossed polarizers) for the spectrometric detection. Error bars represent standard deviations ($n \geq 3$).

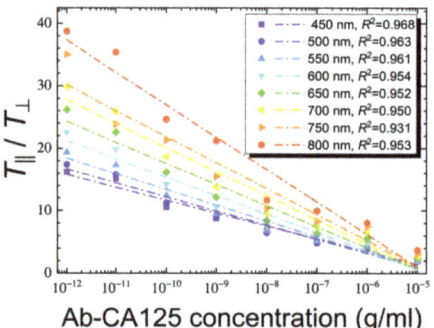

Figure 8. The correlation between the concentration of the anti-CA125 antibody and the ratio of reduced transmittance derived from Figure 7. The $T_\parallel/T_\perp$ ratios were calculated and plotted against the concentration of the anti-CA125 antibody at selected wavelengths of 450, 500, 550, 600, 650, 700, 750, and 800 nm. The coefficient of determination, R^2, for each regression line is given in the legend as a measure of the agreement with the linear fit.

3.4. LCLC-Based Quantitative Immunoassay of the Cancer Biomarker CA125

In the LCLC-based CA125 immunoassay, the biomarker CA125 was captured by the anti-CA125 antibody immobilized at the LC–glass interface through the highly specific antigen–antibody immunoreaction. As demonstrated in the design of our previously reported LC-based CA125 immunoassays, the amount of the immobilized anti-CA125 antibody was optimized so that enough antibodies were present to interact with a wide concentration range of CA125 in the analyte without creating background signals or false-positive results [28–31]. In the LCLC-based immunoassay platform, a maximal concentration was tested and predetermined for the immobilization of the anti-CA125 antibody without interfering with the planar alignment of the LCLC. In accordance with the LOD values at 450 nm for the anti-CA125 antibody, two antibody immobilization concentrations, 10^{-10} and 10^{-9} g/mL, were chosen for comparative studies. As the volume of the antibody solution for immobilization was 30 µL, the total amount of the anti-CA125 antibody reacted with the glass substrate when a 10^{-10} or 10^{-9} g/mL solution was applied corresponds to 3 or 30 pg, respectively. The anti-CA125 antibody was immobilized on a circular area on the PI-coated glass substrate, whose entire surface was then reacted with 10^{-12}–10^{-5} g/mL CA125 so that nonspecific binding to the rubbed PI surface can be easily detected outside the area containing the antibody (Figure 2h–k).

As shown in Figure 9, light leakage caused by the disturbance in the planar alignment of DSCG increased with increasing concentrations of CA125, and it became more significant at low CA125 concentrations when a larger amount of the anti-CA125 antibody was immobilized. The brightness of the optical texture was discernible at 10^{-10} g/mL (10^{-11} g/mL) of CA125 when 10^{-10} g/mL (10^{-9} g/mL) of the anti-CA125 antibody was immobilized, inferring that more CA125 immunocomplexes were formed when 10^{-9} g/mL anti-CA125 antibody was immobilized. When subjected to transmission spectrometric analysis at various CA125 concentrations, the general trend of $T_\parallel$ decreasing and $T_\perp$ increasing with analyte concentration remains (figure not shown), similar to that seen in Figures 5 and 7 for BSA and the anti-CA125 antibody. In general, the LOD obtained at an anti-CA125 antibody concentration of 10^{-9} g/mL (LOD_9) was smaller than that at 10^{-10} g/mL (LOD_{10}), with LOD_9 values obtained from the curves of $T_\parallel$ and $T_\perp$ at 450 nm versus CA125 concentration of 5.0×10^{-11} and 1.1×10^{-10} g/mL, respectively. Actually, at an optimized immobilization concentration of 10^{-9} g/mL anti-CA125 antibody, both the sensitivity and LOD of the LCLC-based CA125 immunoassay were improved in comparison with those obtained at 10^{-10} g/mL anti-CA125 antibody, without compromising the signal-to-noise ratio.

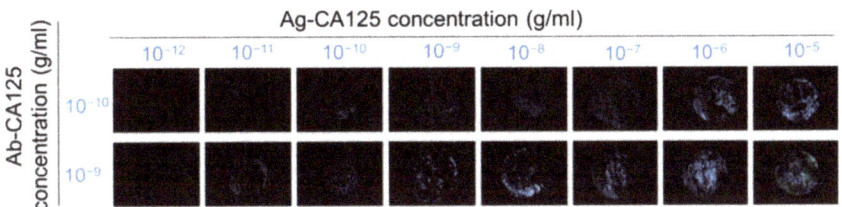

Figure 9. Optical textures of planarly aligned LCLC when the cancer biomarker CA125 was reacted with immobilized anti-CA125 antibody. The LCLC-based immunoassay was performed with an aqueous solution of 14-wt% DSCG as the sensing medium in contact with 10^{-10} or 10^{-9} g/mL anti-CA125 antibody immobilized on the PI-coated glass surface, followed by reaction with 10^{-12}–10^{-5} g/mL CA125.

The ratios of reduced transmittance, $T_\parallel/T_\perp$, were calculated for both antibody immobilization concentrations and were plotted against CA125 concentration at representative wavelengths of 450, 600, and 750 nm, as shown in Figure 10. Similar to Figure 8 for the plots of $T_\parallel/T_\perp$ versus the concentration of the anti-CA125 antibody, a negative and linear correlation was found between the $T_\parallel/T_\perp$ ratio and the CA125 concentration, with the absolute values of $T_\parallel/T_\perp$ and the slope of the linear regression line increased with increasing wavelengths of the incident light (Figure 10). Because the $T_\perp$ values obtained at 10^{-10} g/mL anti-CA125 antibody were generally lower than those at 10^{-9} g/mL anti-CA125 antibody, the absolute values of $T_\parallel/T_\perp$ at each CA125 concentration and the slope of the regression line became larger at a lower antibody immobilization concentration, when each pair of the CA125 calibration curves in Figure 10a–c was compared. These observations imply that, by converting $T_\parallel$ to $T_\parallel/T_\perp$ through data processing, the initially less significant concentration-dependent response in Figure 7c at longer wavelengths or lower antibody immobilization concentrations can be enhanced, as seen in Figures 8 and 10. The calculated LOD_{10} from the linear correlation between $T_\parallel/T_\perp$ and CA125 concentration at the more favorable 800 nm, a condition at which the LOD at an antibody concentration of 10^{-10} g/mL can be more accurately estimated, was 1.7×10^{-10} g/mL CA125.

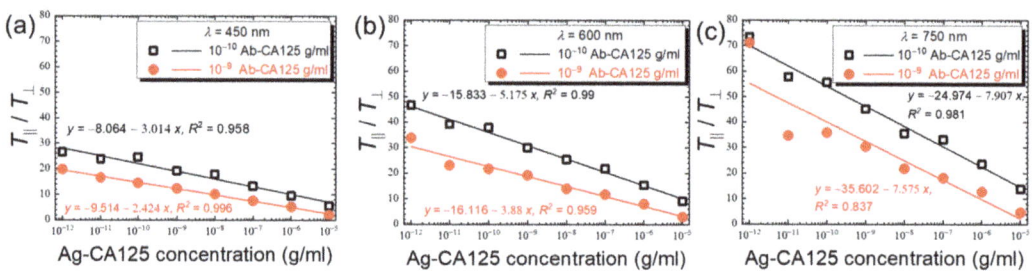

Figure 10. The correlation between CA125 concentration and the ratio of reduced transmittance at two immobilization concentrations of the anti-CA125 antibody, 10^{-10} and 10^{-9} g/mL. The ratios, $T_\parallel/T_\perp$, were calculated and plotted against the concentration of CA125 at selected wavelengths of (**a**) 450, (**b**) 600, and (**c**) 750 nm. The equation and coefficient of determination, R^2, for each linear regression line are shown as a measure of agreement with the linear fit.

In our previous biosensing studies based on thermotropic LCs, the lowest detectable BSA and CA125 concentrations were 10^{-11} g/mL BSA and 10^{-8} g/mL CA125 (with anti-CA125 antibody immobilized at 10^{-7} g/mL), respectively, when a nematic LC of high birefringence was employed as the sensing medium [28]. Through detection with a dye-doped LC in conjunction with transmission spectrometry, analyte concentrations as low as 10^{-6} g/mL BSA and 10^{-5} g/mL CA125 (with anti-CA125 antibody immobilized at 10^{-7} g/mL) can be discerned from the background [31]. In a single-substrate detection platform based on an LC-photopolymer composite film, signal amplification through photopolymerization gave rise to LOD values of 1.6×10^{-12} g/mL BSA and 2.1×10^{-8} g/mL CA125 (with anti-CA125 antibody immobilized at 10^{-10} g/mL), respectively [30]. In this study, the $LOD_\parallel$ and $LOD_\perp$ for BSA at 450 nm were 2.4×10^{-10} and 6.0×10^{-11} g/mL, respectively, whereas those for CA125 were 5.0×10^{-11} and 1.1×10^{-10} g/mL (with anti-CA125 antibody immobilized at 10^{-9} g/mL), respectively. The comparable and even lower LOD achieved by nematic DSCG demonstrates that LCLCs can serve as an alternative to thermotropic LCs as the biosensing media. Moreover, by exploiting the biocompatibility and hydrophilicity of DSCG, the end-point assay demonstrated in this study can be further transformed into real-time detection by mixing the target of detection (e.g., CA125) in the aqueous solution of DSCG and, after injection into the LC cell, monitoring the change in transmittance at a specific wavelength over time as the target protein associates with the immobilized capture molecule (e.g., anti-CA125 antibody).

4. Conclusions

In this study, a quantitative label-free biosensor for BSA protein assay and CA125 immunoassay was developed based on the spectrometric analysis of LCLCs. By employing the nematic phase of aqueous DSCG as the sensing mesogen, the biosensing platform was designed so that when the planar alignment of LCLCs was disrupted by biomolecules present at the LC-glass interface, a concentration- and wavelength-dependent optical signal was produced, which was analyzed by transmission spectrometry in the visible spectrum with parallel or crossed polarizers. The reduced transmittance obtained with parallel or crossed polarizers, $T_\parallel$ or $T_\perp$, was negatively or positively correlated to BSA concentration, respectively, thus enabling calibration curves to be constructed for protein quantitation. The LCLC-based CA125 immunoassay was established with an optimized antibody immobilization concentration of 10^{-9} g/mL anti-CA125 antibody, at which the sensitivity and the LOD were improved compared with those obtained at 10^{-10} g/mL anti-CA125 antibody. In addition, the linear correlation between the $T_\parallel/T_\perp$ ratio and the logarithm of CA125 concentration may offer more accurate quantitative results at longer wavelengths, where the detection sensitivity was lower when only $T_\parallel$ was considered. The results from this study reveal a new perspective on how the nematic phase of LCLCs can be employed in LC-based biosensing similar to thermotropic LCs. By incorporating the water-soluble biomolecular analytes in the hydrophilic LCLCs, it is possible to further endow LCLC-based biosensors with real-time detection capabilities unattainable by hydrophobic thermotropic LCs.

Author Contributions: Conceptualization, H.S., M.-J.L. and W.L.; Methodology, H.S.; Software, H.S.; Validation, H.S., M.-J.L. and W.L.; Formal Analysis, H.S.; Investigation, H.S.; Resources, M.-J.L. and W.L.; Data Curation, H.S.; Writing—Original Draft Preparation, H.S.; Writing—Review and Editing, M.-J.L. and W.L.; Visualization, H.S.; Supervision, M.-J.L. and W.L.; Project Administration, M.-J.L. and W.L.; Funding Acquisition, M.-J.L. and W.L. All authors have read and agreed to the published version of the manuscript.

Funding: This work was financially supported by the Ministry of Science and Technology, Taiwan, under grant Nos. 107-2112-M-009-012-MY3, 109-2320-B-309-001, 110-2112-M-A49-023, and 110-2320-B-309-001.

Institutional Review Board Statement: Not applicable.

Informed Consent Statement: Not applicable.

Data Availability Statement: The authors confirm that the data supporting the findings of this study are available within the article.

Conflicts of Interest: The authors declare no conflict of interest.

References

1. Luan, C.; Luan, H.; Luo, D. Application and technique of liquid crystal-based biosensors. *Micromachines* **2020**, *11*, 176. [CrossRef]
2. Prakash, J.; Parveen, A.; Mishra, Y.K.; Kaushik, A. Nanotechnology-assisted liquid crystals-based biosensors: Towards fundamental to advanced applications. *Biosens. Bioelectron.* **2020**, *168*, 112562. [CrossRef]
3. Wang, Z.; Xu, T.; Noel, A.; Chen, Y.-C.; Liu, T. Applications of liquid crystals in biosensing. *Soft Matter* **2021**, *17*, 4675–4702. [CrossRef]
4. Nayani, K.; Yang, Y.; Yu, H.; Jani, P.; Mavrikakis, M.; Abbott, N. Areas of opportunity related to design of chemical and biological sensors based on liquid crystals. *Liq. Cryst. Today* **2020**, *29*, 24–35. [CrossRef]
5. Popov, P.; Mann, E.K.; Jákli, A. Thermotropic liquid crystal films for biosensors and beyond. *J. Mater. Chem. B* **2017**, *5*, 5061–5078. [CrossRef]
6. Munir, S.; Kang, I.-K.; Park, S.-Y. Polyelectrolytes functionalized nematic liquid crystal-based biosensors: An overview. *TrAC Trends Anal. Chem.* **2016**, *83*, 80–94. [CrossRef]
7. Lee, M.-J.; Lee, W. Liquid crystal-based capacitive, electro-optical and dielectric biosensors for protein quantitation. *Liq. Cryst.* **2020**, *47*, 1145–1153. [CrossRef]
8. Concellón, A.; Fong, D.; Swager, T.M. Complex liquid crystal emulsions for biosensing. *J. Am. Chem. Soc.* **2021**, *143*, 9177–9182. [CrossRef]
9. Wu, P.-C.; Karn, A.; Lee, M.-J.; Lee, W.; Chen, C.-Y. Dye-liquid-crystal-based biosensing for quantitative protein assay. *Dyes Pigment.* **2018**, *150*, 73–78. [CrossRef]
10. Lin, C.-M.; Wu, P.-C.; Lee, M.-J.; Lee, W. Label-free protein quantitation by dielectric spectroscopy of dual-frequency liquid crystal. *Sens. Actuators B Chem.* **2019**, *282*, 158–163. [CrossRef]
11. Lee, M.-J.; Lee, W. *Unconventional Liquid Crystals and Their Applications*; Lee, W., Kumar, S., Eds.; Chapter 5; De Gruyter: Berlin, Germany, 2021; pp. 239–264. [CrossRef]
12. Tam-Chang, S.-W.; Huang, L. Chromonic liquid crystals: Properties and applications as functional materials. *Chem. Commun.* **2008**, *17*, 1957–1967. [CrossRef] [PubMed]
13. Nastishin, Y.A.; Liu, H.; Schneider, T.; Nazarenko, V.; Vasyuta, R.; Shiyanovskii, S.V.; Lavrentovich, O.D. Optical characterization of the nematic lyotropic chromonic liquid crystals: Light absorption, birefringence, and scalar order parameter. *Phys. Rev. E Stat. Nonlinear Soft Matter Phys.* **2005**, *72*, 41711. [CrossRef] [PubMed]

14. Agra-Kooijman, D.M.; Singh, G.; Lorenz, A.; Collings, P.J.; Kitzerow, H.S.; Kumar, S. Columnar molecular aggregation in the aqueous solutions of disodium cromoglycate. *Phys. Rev. E Stat. Nonlinear Soft Matter Phys.* **2014**, *89*, 062504. [CrossRef]
15. Luk, Y.Y.; Jang, C.H.; Cheng, L.L.; Israel, B.A.; Abbott, N.L. Influence of lyotropic liquid crystals on the ability of antibodies to bind to surface-immobilized antigens. *Chem. Mater.* **2005**, *17*, 4774–4782. [CrossRef]
16. Lydon, J. Chromonic review. *J. Mater. Chem.* **2010**, *20*, 10071–10099. [CrossRef]
17. Dhakal, N.P.; Jiang, J.; Guo, Y.; Peng, C. Self-assembly of aqueous soft matter patterned by liquid-crystal polymer networks for controlling the dynamics of bacteria. *ACS Appl. Mater. Inter.* **2020**, *12*, 13680–13685. [CrossRef]
18. Kumar, A.; Galstian, T.; Pattanayek, S.K.; Rainville, S. The motility of bacteria in an anisotropic liquid environment. *Mol. Cryst. Liq. Cryst.* **2013**, *574*, 33–39. [CrossRef]
19. Mushenheim, P.C.; Trivedi, R.R.; Weibel, D.B.; Abbott, N.L. Using liquid crystals to reveal how mechanical anisotropy changes interfacial behaviors of motile bacteria. *Biophys. J.* **2014**, *107*, 255–265. [CrossRef]
20. Zhou, S.; Tovkach, O.; Golovaty, D.; Sokolov, A.; Aranson, I.S.; Lavrentovich, O.D. Dynamic states of swimming bacteria in a nematic liquid crystal cell with homeotropic alignment. *New J. Phys.* **2017**, *19*, 055006. [CrossRef]
21. Cheng, L.L.; Luk, Y.Y.; Murphy, C.J.; Israel, B.A.; Abbott, N.L. Compatibility of lyotropic liquid crystals with viruses and mammalian cells that support the replication of viruses. *Biomaterials* **2005**, *26*, 7173–7182. [CrossRef]
22. Shiyanovskii, S.V.; Lavrentovich, O.D.; Schneider, T.; Ishikawa, T.; Smalyukh, I.I.; Woolverton, C.J.; Niehaus, G.D.; Doane, K.J. Lyotropic chromonic liquid crystals for biological sensing applications. *Mol. Cryst. Liq. Cryst.* **2005**, *434*, 259–270. [CrossRef]
23. Skates, S.J.; Xu, F.-J.; Yu, Y.-H.; Sjövall, K.; Einhorn, N.; Chang, Y.; Bast, R.C.; Knapp, R.C. Toward an optimal algorithm for ovarian cancer screening with longitudinal tumor markers. *Cancer* **1995**, *76*, 2004–2010. [CrossRef]
24. Cotchim, S.; Thavarungkul, P.; Kanatharana, P.; Limbut, W. Multiplexed label-free electrochemical immunosensor for breast cancer precision medicine. *Anal. Chim. Acta* **2020**, *1130*, 60–71. [CrossRef]
25. de Castro, A.C.H.; Alves, L.M.; Siquieroli, A.C.S.; Madurro, J.M.; Brito-Madurro, A.G. Label-free electrochemical immunosensor for detection of oncomarker CA125 in serum. *Microchem. J.* **2020**, *155*, 104746. [CrossRef]
26. Mansouri Majd, S.; Salimi, A. Ultrasensitive flexible FET-type aptasensor for CA 125 cancer marker detection based on carboxylated multiwalled carbon nanotubes immobilized onto reduced graphene oxide film. *Anal. Chim. Acta* **2018**, *1000*, 273–282. [CrossRef] [PubMed]
27. Ravalli, A.; Pilon Dos Santos, G.; Ferroni, M.; Faglia, G.; Yamanaka, H.; Marrazza, G. New label free CA125 detection based on gold nanostructured screen-printed electrode. *Sens. Actuat. B Chem.* **2013**, *179*, 194–200. [CrossRef]
28. Su, H.-W.; Lee, Y.-H.; Lee, M.-J.; Hsu, Y.-C.; Lee, W. Label-free immunodetection of the cancer biomarker CA125 using high-Δn liquid crystals. *J. Biomed. Opt.* **2014**, *19*, 077006. [CrossRef] [PubMed]
29. Su, H.-W.; Lee, M.-J.; Lee, W. Surface modification of alignment layer by ultraviolet irradiation to dramatically improve the detection limit of liquid-crystal-based immunoassay for the cancer biomarker CA125. *J. Biomed. Opt.* **2015**, *20*, 057004. [CrossRef]
30. Lee, M.-J.; Duan, F.-F.; Wu, P.-C.; Lee, W. Liquid crystal–photopolymer composite films for label-free single-substrate protein quantitation and immunoassay. *Biomed. Opt. Express* **2020**, *11*, 4915. [CrossRef]
31. Chiang, Y.L.; Lee, M.J.; Lee, W. Enhancing detection sensitivity in quantitative protein detection based on dye-doped liquid crystals. *Dyes Pigment.* **2018**, *157*, 117–122. [CrossRef]
32. Yang, K.H. Measurements of empty cell gap for liquid-crystal displays using interferometric methods. *J. Appl. Phys.* **1988**, *64*, 4780–4781. [CrossRef]
33. Zhou, S.; Neupane, K.; Nastishin, Y.A.; Baldwin, A.R.; Shiyanovskii, S.V.; Lavrentovich, O.D.; Sprunt, S. Elasticity, viscosity, and orientational fluctuations of a lyotropic chromonic nematic liquid crystal disodium cromoglycate. *Soft Matter* **2014**, *10*, 6571–6581. [CrossRef]
34. Kumar, A.; Pattanayek, S.K. Exploitation of orientation of liquid crystals 5CB and DSCG near surfaces to detect low protein concentration. *Liq. Cryst.* **2015**, *42*, 1506–1514. [CrossRef]
35. Nazarenko, V.G.; Boiko, O.P.; Park, H.S.; Brodyn, O.M.; Omelchenko, M.M.; Tortora, L.; Nastishin, Y.A.; Lavrentovich, O.D. Surface alignment and anchoring transitions in nematic lyotropic chromonic liquid crystal. *Phys. Rev. Lett.* **2010**, *105*, 1–4. [CrossRef]
36. Guo, Y.; Shahsavan, H.; Davidson, Z.S.; Sitti, M. Precise Control of Lyotropic Chromonic Liquid Crystal Alignment through Surface Topography. *ACS Appl. Mater. Interfaces* **2019**, *11*, 36110–36117. [CrossRef] [PubMed]
37. Jeong, J.; Han, G.; Johnson, A.T.C.; Collings, P.J.; Lubensky, T.C.; Yodh, A.G. Homeotropic alignment of lyotropic chromonic liquid crystals using noncovalent interactions. *Langmuir* **2014**, *30*, 2914–2920. [CrossRef]
38. Collings, P.J.; van der Asdonk, P.; Martinez, A.; Tortora, L.; Kouwer, P.H.J. Anchoring strength measurements of a lyotropic chromonic liquid crystal on rubbed polyimide surfaces. *Liq. Cryst.* **2017**, *44*, 1165–1172. [CrossRef]
39. Shrivastava, A.; Gupta, V. Methods for the determination of limit of detection and limit of quantitation of the analytical methods. *Chron. Young Sci.* **2011**, *2*, 21. [CrossRef]
40. Hu, Y.J.; Liu, Y.; Sun, T.Q.; Bai, A.M.; Lü, J.Q.; Pi, Z.B. Binding of anti-inflammatory drug cromolyn sodium to bovine serum albumin. *Int. J. Biol. Macromol.* **2006**, *39*, 280–285. [CrossRef]

Article

Signal Amplification in an Optical and Dielectric Biosensor Employing Liquid Crystal-Photopolymer Composite as the Sensing Medium

Hassanein Shaban [1,2], Shih-Chun Yen [1], Mon-Juan Lee [3,4,*] and Wei Lee [1,*]

1. Institute of Imaging and Biomedical Photonics, College of Photonics, National Yang Ming Chiao Tung University, Guiren District, Tainan 71150, Taiwan; hassanein.shaban@sci.asu.edu.eg (H.S.); loppet3@gmail.com (S.-C.Y.)
2. Department of Basic Science, Faculty of Engineering, The British University in Egypt, El Sherouk City 11837, Cairo, Egypt
3. Department of Bioscience Technology, Chang Jung Christian University, Guiren District, Tainan 71101, Taiwan
4. Department of Medical Science Industries, Chang Jung Christian University, Guiren District, Tainan 71101, Taiwan
* Correspondence: mjlee@mail.cjcu.edu.tw (M.-J.L.); wlee@nctu.edu.tw (W.L.)

Citation: Shaban, H.; Yen, S.-C.; Lee, M.-J.; Lee, W. Signal Amplification in an Optical and Dielectric Biosensor Employing Liquid Crystal-Photopolymer Composite as the Sensing Medium. *Biosensors* **2021**, *11*, 81. https://doi.org/10.3390/bios11030081

Received: 2 February 2021
Accepted: 11 March 2021
Published: 13 March 2021

Publisher's Note: MDPI stays neutral with regard to jurisdictional claims in published maps and institutional affiliations.

Copyright: © 2021 by the authors. Licensee MDPI, Basel, Switzerland. This article is an open access article distributed under the terms and conditions of the Creative Commons Attribution (CC BY) license (https://creativecommons.org/licenses/by/4.0/).

Abstract: An optical and dielectric biosensor based on a liquid crystal (LC)–photopolymer composite was established in this study for the detection and quantitation of bovine serum albumin (BSA). When the nematic LC E7 was doped with 4-wt.% NOA65, a photo-curable prepolymer, and photopolymerized by UV irradiation at 20 mW/cm^2 for 300 s, the limit of detection determined by image analysis of the LC optical texture and dielectric spectroscopic measurements was 3400 and 88 pg/mL for BSA, respectively, which were lower than those detected with E7 alone (10 µg/mL BSA). The photopolymerized NOA65, but not the prepolymer prior to UV exposure, contributed to the enhanced optical signal, and UV irradiation of pristine E7 in the absence of NOA65 had no effect on the optical texture. The effective tilt angle θ, calculated from the real-part dielectric constant ε', decreased with increasing BSA concentration, providing strong evidence for the correlation of photopolymerized NOA65 to the intensified disruption in the vertically oriented LC molecules to enhance the optical and dielectric signals of BSA. The optical and dielectric anisotropy of LCs and the photo-curable dopant facilitate novel quantitative and signal amplification approaches to potential development of LC-based biosensors.

Keywords: liquid crystal; photopolymer; UV exposure; bovine serum albumin; protein assay; dielectric spectroscopy

1. Introduction

Biosensors are devices designed for the detection of biologically relevant small molecules, biomolecules, biomolecular interactions such as the binding between antigen and antibody, or whole cells such as bacteria and viruses. The biological signals produced by the target of detection are transduced through the biosensor into electrical, thermal or optical signals for further qualitative and quantitative data processing. Liquid crystals (LCs) are considered novel biosensing media because of their sensitive optical response to biomolecules, thus enabling label-free bioassays based on LCs to be established. LC-based biosensing technologies can be subdivided into two platforms, one of which relies on detection at the LC-water interface, whereas the other employs the LC-glass interface.

Biosensing at the LC-water interface was applied in the detection of proteins [1], lipids [2,3], amphiphilic molecules [4,5], DNA [6], enzymatic activity [7,8] and immunocomplexes [9]. This design is characterized by the capacity for real-time detection as biological materials are water-soluble, but the optical signal derived from the LC texture can only provide qualitative or semiquantitative results. On the other hand, biodetection at the LC-glass interface is categorized as an end-point assay, where the biorecognition and biomolecular reactions are completed on a glass substrate prior to LC cell assembly with another glass substrate. In fact, sandwiching LCs between two glass substrates as an LC cell enables the external application of an electric field so that not only the optical anisotropy but also the electrical and electro-optical properties of LCs can be utilized in biosensing [10]. By exploring various LCs other than the narrow-nematic-range 5CB used in most LC-based biosensors, we developed a quantitative protein assay and immunoassay in conjunction with transmission spectrometry [11–14], as well as LC-based capacitive [15], electro-optical [14] and dielectric biosensors [16].

One of the major technical hurdles of such label-free biosensing techniques is the limited approaches for signal amplification. Gold nanoparticles were reported in several studies to enhance the optical signal of LCs by forming complexes with the target of detection or by altering the surface topology of the sensing interface to increase the extent of disturbance in the ordered alignment of LCs [17–19]. Our previous work demonstrated that by using LCs of high birefringence [20–22] or by exploiting the electrically inducible potential of LCs [23], the optical signal in LC-based biosensors can be promoted. Modification of the LC alignment layer—say, the surfactant dimethyloctadecyl[3-(trimethoxysilyl)propyl]ammonium chloride (DMOAP), a common vertical alignment reagent—coated on the glass substrate with ultraviolet (UV) irradiation [20] and adjusting the polarization direction of linearly polarized light in accordance with the rubbing direction of the glass substrate [13] also contribute to signal amplification.

Various non-LC materials, such as dichroic dyes, chiral compounds and nanomaterials, can be incorporated into LCs, giving rise to composite materials such as dye-doped LCs and cholesteric LCs that exhibit unique characteristics unattainable with pristine LCs. We previously reported that the optical signal derived from a single-substrate biosensor based on LC-photopolymer composites can be enhanced by fine-tuning the level of photopolymerization of the dopant, the photo-curable NOA65 prepolymer [24]. Studies have shown that when a mixture of NOA65 and the nematic LC E7 was exposed to UV, the vertical phase separation due to the difference in surface tension between NOA65 and E7 led to the accumulation and polymerization of NOA65 at the LC-glass interface [25,26]. The gravel-like NOA65 photopolymer thus increased the roughness and polarity of the glass surface, and was therefore exploited to control the pretilt angle of the LC molecules. When the concentration of NOA65 doped in E7 was increased from 0 wt.% to 2.5 wt.%, the pretilt angle of E7 reduced from 87.3° to 2.5° [25].

To further our understanding of the mechanism of signal amplification provided by photopolymerized NOA65, an optical and dielectric biosensing system based on the UV-cured NOA65/E7 composite was established in this study for the detection and quantitation of bovine serum albumin (BSA), a common calibration standard for the determination of protein concentrations in biological analytes. Instead of the single-substrate platform constructed in our previous work [24], the NOA65/E7 composite was sandwiched between two parallel glass substrates as a LC cell to facilitate the application of an electric field for dielectric analysis. Optical and dielectric measurement were performed on the NOA65/E7 composites at various NOA65 concentrations, UV intensities and exposure times to study the effect of the level of photopolymerization on signal amplification. The dielectric anisotropy of LCs and results derived from dielectric measurements offered a new approach to study the effect of photopolymerized NOA65 on LC orientation and signal amplification.

2. Experimental
2.1. Materials

Optical glass substrates with dimensions 22 mm × 18 mm × 1.1 mm were obtained from Ruilong Glass, Miaoli, Taiwan. Indium–tin-oxide (ITO)-coated conductive glass slides, a pair of which produces an overlapped electrode area of 5.0 mm × 5.0 mm, were manufactured by Chipset Technology Co., Ltd., Miaoli, Taiwan. The nematic LC E7 used in this study is produced by Daily Polymer Corp., Kaohsiung, Taiwan. The birefringence Δn

of E7 at a wavelength of 589 nm and a temperature of 20 °C is 0.2255, with the real part of the dielectric constant parallel and perpendicular to the LC molecular axis, $\varepsilon_\parallel$ = 19.5 and $\varepsilon_\perp$ = 5.2, respectively, at a frequency of 1 kHz. The photo-curable prepolymer NOA65, which is an adhesive commonly included in polymer-dispersed LCs, was obtained from Norland Products, Inc., Cranbury, NJ, USA. Vertical alignment of LCs was achieved with DMOAP, which was purchased from Sigma–Aldrich, St. Louis, MO, USA. DMOAP self-assembled into a monolayer on a glass substrate and effectively aligned LC molecules in the direction of its long alkyl chain, $-CH_3(CH_2)_{16}CH_2$. BSA, a conventional protein standard consisting of 583 amino acid residues with a molecular weight of approximately 66 kDa, was provided by Sigma–Aldrich, St. Louis, MO, USA.

2.2. Preparation of DMOAP-Coated Glass Substrates

Steps for preparing substrates used in the cell platform for biological detection are shown in Figure 1. Optical or ITO glass slides were first cleaned by sonication in a detergent solution and then washed twice in deionized (DI) water and once in ethanol, with each procedure lasted for 15 min with sonication (Figure 1a). The substrates were dried with nitrogen and then baked in an oven at 74 °C for 30 min. Each cleaned substrate was dipped under the application of ultrasound in a 0.1% (*v/v*) DMOAP solution for 15 min, and then washed twice with DI water for 15 min to facilitate self-assembly of the monolayer (Figure 1b). The dip-coated substrate was blown with nitrogen and heated in the oven at 85 °C for 15 min to cure the aligning monolayer for imposing vertical alignment of LC molecules.

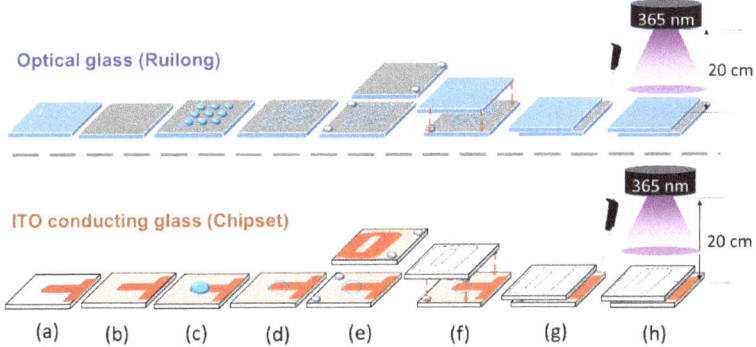

Figure 1. Procedures for establishing the biosensing platform based on the liquid crystal (LC)–photopolymer composite in which LC cells were assembled with either optical glass or indium–tin-oxide (ITO) conductive glass for optical or dielectric analysis, respectively. The glass substrates were cleaned to remove contaminants (**a**), followed by coating with dimethyloctadecyl[3-(trimethoxysilyl)propyl]ammonium chloride (DMOAP) as the vertical alignment layer (**b**) and immobilization with bovine serum albumin (BSA) (**c**). After drying at 35 °C on a hot plate (**d**), a mixture of 5.5 μm ball spacer and AB glue was applied at the corners of a pair of substrates (**e**) for the assembly of the LC cell (**f**). The NOA65/E7 mixture was then injected with a micropipette (**g**) and irradiated with UV light at irradiance of 5–20 mW/cm^2 (**h**).

2.3. Fabrication of the LC Cell and Immobilization of BSA Molecules

BSA solutions of concentrations ranging from 10^{-12} to 10^{-5} g/mL were prepared in DI water. For BSA immobilization on a DMOAP-coated optical glass substrate, a 3 × 3 protein array was formed with 3 μL BSA solution per spot (Figure 1c). For the DMOAP-coated ITO glass substrate, a 33-μL BSA solution was dispensed to cover the entire 0.25-cm^2 electrode area. The BSA solution on the glass substrates was dried for 20 min on a hot plate with the temperature set at 35 °C (Figure 1d). 5.5-μm ball spacers and an AB glue were mixed and dispensed on the two corners of the upper substrate (without BSA) and the lower substrate (with immobilized BSA), respectively, as shown in Figure 1e. LC cell assembly was performed by gently pressing the pair of glass substrates together and allowing the AB glue to dry for 30 min (Figure 1f). The cell gap of the assembled LC cell was measured by optical interferometry with an Ocean Optics HR2000+ high-resolution USB fiber-optic spectrometer [27]. Each LC cell was then filled with a mixture of E7 and NOA65 through capillary action by injecting the mixture with a micropipette from the side of the LC cell (Figure 1g), followed by UV exposure at wavelength of 365 nm with a Panasonic Aicure UJ35 LED Spot Type UV Curing System to induce photopolymerization (Figure 1h).

2.4. Optical Measurement and Image Analysis with the ImageJ Software

An Olympus BX51-P polarizing optical microscope (POM) with crossed polarizers in the transmission mode was employed for the observation of LC textures, and images with a resolution of 2048 × 1536 pixels were taken with an Olympus XC30 digital camera. To perform quantitative analysis, ImageJ, an open-source image processing and analysis program, was used to determine the relative intensity of each optical texture image by averaging the brightness of the three primary colors of RGB (0–255) with the formula $V = (R + G + B)/3$.

2.5. Dielectric Measurement

The real part of the dielectric constant was measured by a Hioki 3522-50 LCR meter, which was interfaced with a computer through a GPIB interface card and the LabVIEW graphic control program. A probe AC voltage of no more than 0.1 V was applied, which was lower than the transition threshold voltage of E7 within a frequency range of 10 Hz to 100 kHz. The real part of the dielectric spectra at various BSA concentration was recorded, and the dielectric constant at a frequency of 1 kHz was used in protein quantitative analysis.

3. Results and Discussion

3.1. Biosensing Based on the Optical Measurement of the LC–Photopolymer Composite

LC-based biosensing is facilitated by the interaction between biomolecules and LCs, whose orientation responds sensitively to the disturbance caused by the analyte, thereby generating optical and electro-optical signals that are proportional to the amount of biomolecules. The LC molecules are aligned homeotropically on a substrate coated with a vertical alignment film (DMOAP), giving rise to a dark state when observed under the POM. When biomolecules (BSA) were immobilized on the DMOAP-coated substrate, the vertical anchoring energy of DMOAP was weakened so that the LC molecules were more inclined to arrange randomly, resulting in light leakage and a dark-to-bright transition in the optical texture. In general, an enhanced optical response is considered to be correlated to an increase in the amount of the analyte located at the LC-glass interface. To improve the sensitivity and limit of detection, a photo-curable prepolymer NOA65 was added to the nematic E7 in this study to further amplify the optical signal.

As shown in Figure 2a, in the absence of analytes the vertical anchoring strength of DMOAP was unaffected when a minute concentration of the NOA65 prepolymer was dispersed in E7. After UV exposure, the NOA65 prepolymer phase-separated and aggregated as small polymer gravels on the glass substrate due to the difference in surface tension between NOA65 and E7 [25,26], but the tilt angle of LCs remained unchanged (Figure 2b). In the presence of a trace number of biomolecules, the orientation of LCs was slightly disturbed and the alignment effect of DMOAP was masked by the analyte, but because of the strong vertical anchoring exerted by DMOAP from both glass substrates, optical signals derived from light leakage was undetectable (Figure 2c). Nevertheless, when photopolymerization of NOA65 was induced by UV exposure, surface roughness on the glass substrate was higher in the area with the immobilized biomolecules than that with DMOAP modification only, leading to greater disruption in LC orientation and, in turn, an enhanced optical signal (Figure 2d). It is assumed that the ionic and polar side chains of amino acids distributed on the surface of a protein analyte (BSA in this study) attracted the relatively hydrophilic NOA65, which coalesced around BSA while being repelled by the hydrophobic alkyl chain of DMOAP, bringing about the difference in surface roughness in a situation similar to the preparation of self-positioning NOA65 micro lens [28]. In addition, light scattering can be attributed to refractive-index mismatch in the multi-regional boundaries between the LC molecules, photopolymerized NOA65, DMOAP, BSA, and the glass substrate. As a consequence, LC molecules in the proximity of BSA were disturbed to a greater extent compared with those in direct contact with DMOAP, contributing to signal amplification of BSA without simultaneously increasing the background.

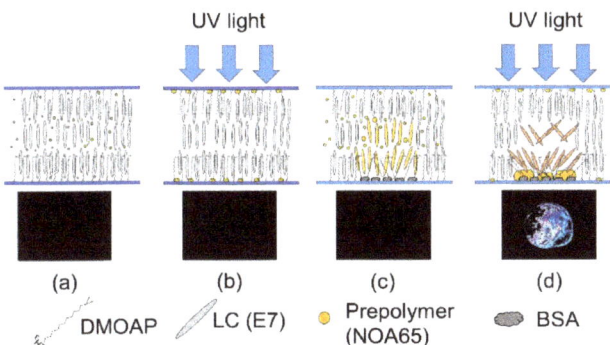

Figure 2. The working principle of the biosensing platform based on the LC–photopolymer composite. (**a**) Before UV exposure and in the absence of biomolecules, the LC molecules are vertically aligned in the presence of a minute concentration of NOA65, resulting in a completely dark optical texture. (**b**) After UV exposure, polymer gravels of NOA65 form on the substrates but are still unable to weaken the vertical anchoring of DMOAP. (**c**) In the presence of immobilized biomolecules at relatively low concentrations, the optical texture is still dark, which implies that the tilt angle of the LC molecules was not significantly affected. (**d**) To achieve signal amplification, the mixture of LC and NOA65 was irradiated with UV to induce the polymerization of NOA65, which leads to significant change in the tilt angle of LC molecules, giving rise to enhanced brightness in the optical texture.

3.1.1. Experimental Conditions for the Preparation of the LC–Photopolymer Composite

To avoid false-positive optical signals, the extent of photopolymerization of NOA65 was carefully controlled so that in the absence of BSA the LC molecules remained vertically anchored and a dark background was observed for the LC–photopolymer composite under the POM with crossed polarizers. As a protein-free reference, DI water instead of BSA was used as the sample, which was dispensed and dried on the glass substrate, followed by LC cell assembly and interaction with the mixture of E7 and NOA65, as described in Section 2.3 and Figure 1. As shown in the upper panels of Figure 3a, when E7 was doped with 1, 2, 4, 5, 7 or 10 wt.% of NOA65, the optical texture of the NOA65/E7 mixture remained dark. After irradiated with UV at 10 mW/cm^2 for 30 s, light leakage was observed at 5-, 7- and 10-wt.% NOA65 (lower panels, Figure 3a). NOA65 concentration was thus maintained at 4 wt.% or lower in the following studies to avoid such nonspecific background noise. We then increased the UV intensity to 20 mW/cm^2 and prolonged the exposure time to 300 s for E7 doped with 4 or 5 wt.% of NOA65 (Figure 3b). A pronounced light leakage was detected when the NOA65/E7 mixture containing 5-wt.% NOA65 was irradiated with UV for 30 s, whereas at 4 wt.% NOA65 the optical texture was completely dark up to a UV exposure time of 300 s. It can be concluded from the above results that in the absence of analytes at the LC-glass interface, LCs remained homeotropically aligned when NOA65 of concentrations ≤4 wt.% was exposed to a UV irradiance of no more than 20 mW/cm^2 for a period of time ≤300 s.

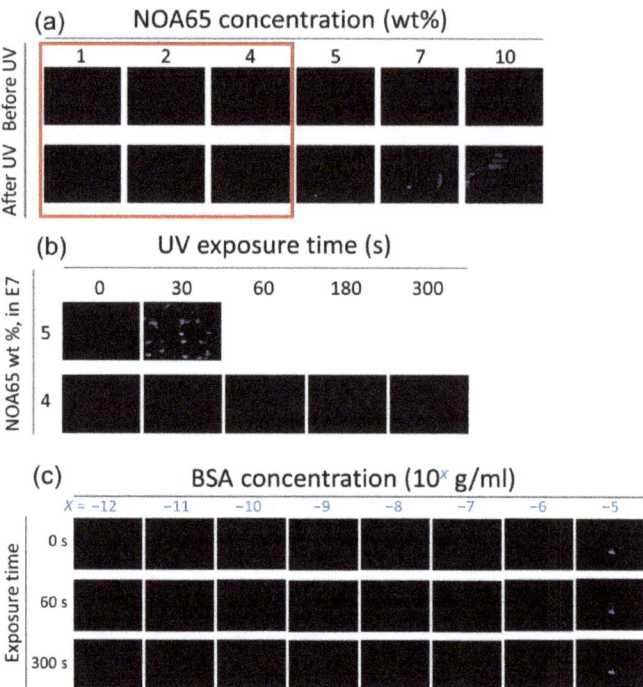

Figure 3. The optical texture of the LC–photopolymer composites at various NOA65 concentrations and UV exposure times. (**a**) E7 was doped with 1, 2, 4, 5, 7, or 10 wt.% NOA65 followed by exposure to 10-mW/cm^2 UV light for 30 s. (**b**) E7 was doped with 4 or 5 wt.% NOA65 followed by exposure to 20-mW/cm^2 UV light for 0, 30, 60, 180, or 300 s. (**c**) The optical texture of pristine E7 in the presence of immobilized BSA molecules. BSA was immobilized at concentrations ranging from 10^{-12} to 10^{-5} g/mL, followed by LC cell assembly and UV irradiation at 20 mW/cm^2 for 0, 60, or 300 s.

3.1.2. Protein Detection and Quantitation with the LC–Photopolymer Composite

To demonstrate signal amplification by photopolymerized NOA65, detection of BSA with pristine E7 as well as the NOA65/E7 composite was compared. As shown in Figure 3c, LC cells were assembled with immobilized BSA in the concentration range of 10^{-12} to 10^{-5} g/mL on one of the DMOAP-coated glass substrates, followed by injection of only E7 and UV irradiation at 20 mW/cm^2 for 0, 60 or 300 s. Without NOA65 the lowest BSA concentration that can be discerned was 10^{-5} g/mL from the optical texture of E7 under the POM. The optical signal at each BSA concentration remained unchanged when irradiated with UV for 60 or 300 s, suggesting that exposure to UV had no effect on the orientation of E7 itself. On the other hand, when doped with 1-wt.% NOA65, the optical texture of the NOA65/E7 mixture in the presence of BSA before UV irradiation was completely dark, similar to that of pristine E7 (Figure 3c). When the LC cell was exposed to UV for 180 s at an intensity of 5, 10 or 20 mW/cm^2, optical signals can be observed at BSA concentrations lower than 10^{-5} g/mL, and the brightness of the optical texture increased with increasing amount of BSA as well as UV intensity (Figure 4a–c), suggesting that the increase in surface roughness caused by photopolymerization of NOA65, as explained in Figure 2, was responsible for the enhanced light leakage and amplified optical signal. In order to quantitatively analyze the optical signal in relation to BSA concentration, the freeware ImageJ was used to perform image analysis and calculate the relative brightness of the optical texture. Results from the quantitative analysis, presented as a plot of relative intensity versus BSA concentration in Figure 4d, was consistent with the texture observations (Figure 4a–c). It was found that the higher the BSA concentration, the greater the enhancement in the texture brightness of the

NOA65/E7 composite with UV intensities. When exposed to 20-mW/cm² UV, the texture brightness of the NOA65/E7 composite was significantly higher than that exposed to 5- or 10-mW/cm² UV in the higher BSA concentration range of 10^{-10}–10^{-5} g/mL, but not at the lower 10^{-12} and 10^{-11} g/mL BSA concentrations. The limit of detection (LOD) was calculated according to the following equation:

$$\text{LOD} = \frac{3s}{m} \tag{1}$$

where s represents the standard deviation of the relative intensity (texture brightness) at the lowest BSA concentration, with its value significantly higher than that at 0-g/mL BSA, and m represents the slope of the linear regression [29]. Based on the results in Figure 4d, the LOD values thus obtained were 1.0×10^{-8}, 8.8×10^{-8} and 6.1×10^{-9} g/mL BSA, when the NOA65/E7 mixture was exposed to UV irradiation of 5, 10 and 20 mW/cm², respectively. The calculated LOD was consistent with the BSA concentration at which the dark-to-bright transition was observed in Figure 4a–c.

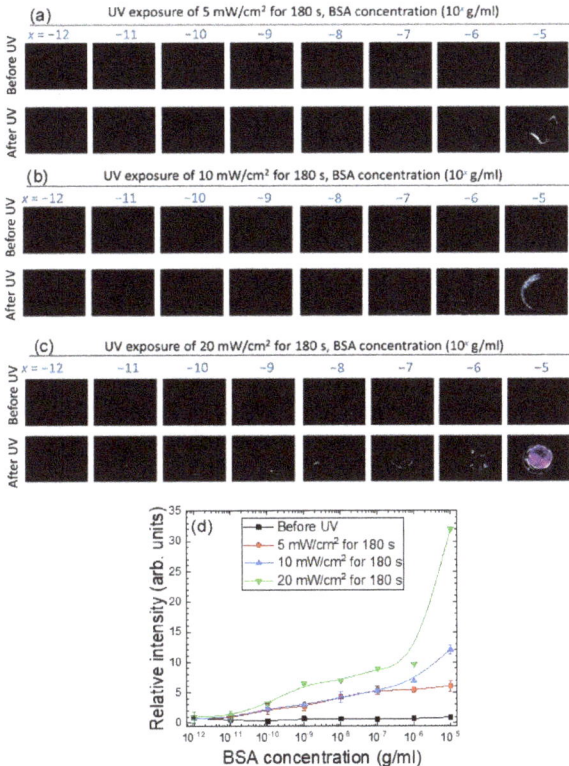

Figure 4. The optical texture of E7 doped with 1-wt.% NOA65 in the presence of immobilized BSA molecules at concentrations ranging from 10^{-12} to 10^{-5} g/mL, followed by LC cell assembly and UV irradiation at (**a**) 5, (**b**) 10 or (**c**) 20 mW/cm² for 180 s. The brightness of the optical textures in (**a**–**c**) was quantitated with ImageJ and plotted against the BSA concentration in (**d**). Error bars represent standard deviations calculated from the relative intensities of at least three independent experiments.

We next increased the doping concentration of NOA65 to 4 wt.%, the maximal concentration at which the homeotropic alignment of E7 can still be maintained without being significantly interfered by the photopolymerized NOA65 (Figure 3). The NOA65/E7 mixture was exposed to UV at an irradiance of 5, 10 or 20 mW/cm² for 180 s (Figure 5a) or for 60, 180 or 300 s at a fixed UV irradiance of 20 mW/cm^2 (Figure 5b). As expected, the relative intensity of the optical texture of the NOA65/E7 composite increased with increasing BSA concentration, and was further enhanced by increasing UV irradiance and exposure time, except for those at 10^{-5}-g/mL BSA, where the brightness of the optical texture seemed to reach saturation and no longer correlated to UV irradiance (Figure 5c,d). The LOD values for the detection by E7 doped with 4 wt.% NOA65 and exposed to UV at 20 mW/cm^2 for 180 and 300 s were 4.3×10^{-8} and 3.4×10^{-9} g/mL BSA, respectively. Compared with 1-wt.% NOA65, the optical signal was significantly amplified by doping E7 with 4-wt.% NOA65, especially at higher concentrations (10^{-8} to 10^{-5} g/mL) of BSA (compare Figures 4d and 5c). However, at a fixed NOA65 concentration, increasing UV irradiance or prolonging UV exposure rendered relatively limited signal amplification (Figures 4d and 5c,d).

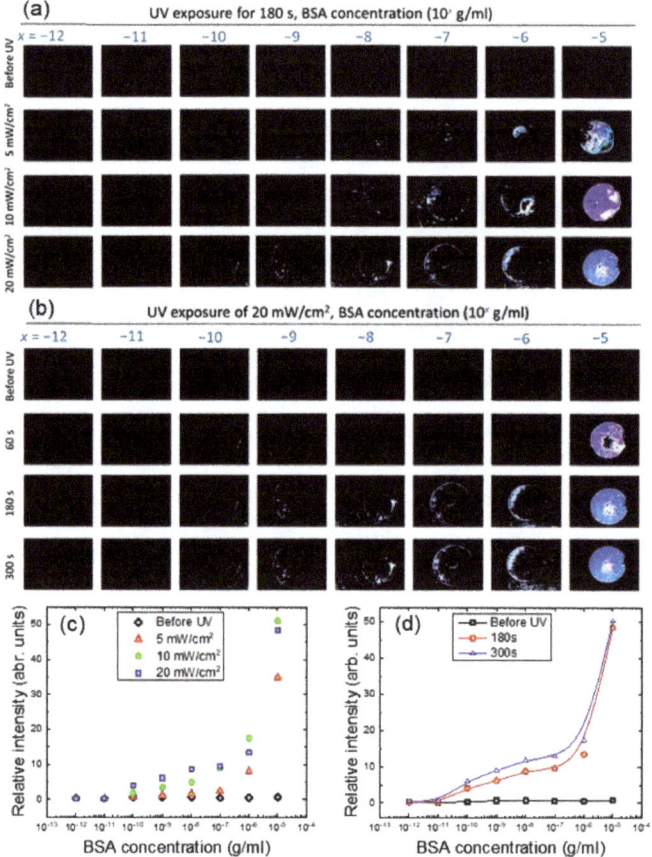

Figure 5. The optical texture of E7 doped with 4-wt.% NOA65 in the presence of BSA molecules. BSA was immobilized at concentrations ranging from 10^{-12} to 10^{-5} g/mL, followed by LC cell assembly and UV irradiation (**a**) at 5, 10, or 20 mW/cm^2 for 180 s, or (**b**) at 20 mW/cm^2 for 0, 60, 180, or 300 s. The brightness of the optical textures in Figure (**a,b**) was quantitated with ImageJ and plotted against the BSA concentration as depicted in Figure (**c,d**), respectively.

In one of our previous studies, a single-substrate biodetection platform was constructed by spin-coating a thin layer of the NOA65/E7 composite film on a DMOAP-modified glass substrate, thus eliminating the procedure for LC cell assembly [24]. Since the thickness of the LC thin film (4.5 ± 0.5 µm) was less than the cell gap of the LC cell (5.5 ± 0.5 µm) assembled in this study, and the vertical anchoring strength at the LC-air interface was much weaker in comparison with that imposed by DMOAP at the LC-glass interface, it is predictable that the LC thin film layered on a single glass substrate is more sensitive to the disturbance caused by biomolecules. The LOD determined for the single-substrate detection was 1.8×10^{-9} g/mL of BSA probed with a NOA65/E7 composite film containing 3 wt.% of NOA65 photopolymerized by UV irradiation at 10 mW/cm^2 for 30 s [24]. Although assembly of an LC cell increased the complexity of the biosensing procedure, the LC film sandwiched between two parallel glass substrates was more stable and uniform in thickness. Besides, by increasing the NOA65 concentration to 4 wt.% and UV exposure to 20 mW/cm^2 for 300 s, a similar LOD of 3.4×10^{-9} g/mL BSA can be achieved with the cell-based biosensing platform.

3.2. Biosensing Based on the Dielectric Measurement of the LC–Photopolymer Composite

For dielectric measurements, the LC cell was assembled with a pair of conductive ITO glass substrates instead of the optical glass used in optical measurements (Figure 1). As a comparison, the transmittance of ITO-coated glass substrates was lower than that of optical flat glass at 365 nm, which is the central wavelength of the UV source for photoinduced polymerization (Figure 6a). LC cells assembled with the ITO-coated glass substrates allowed for an electric field to be applied to measure the capacitance of the NOA65/E7 composite at various BSA concentrations, from which the real part of the dielectric constant was derived. The effective dielectric constant depends on the average tilt angle θ (measured from the substrate plane) of the LC molecules,

$$\varepsilon_{\text{eff}} = \varepsilon_\parallel \sin^2 \theta + \varepsilon_\perp \cos^2 \theta \tag{2}$$

where ε_{eff} represents the measured real-part dielectric constant ε', and $\varepsilon_\parallel$ and $\varepsilon_\perp$ are the parallel and perpendicular components of ε', respectively [15]. It is reasoned that the immobilized BSA may disturb the ordered alignment of LCs, thus altering their tilt angle and consequently the measured dielectric constant. Because surface roughness was increased due to the accumulation of the photopolymerized NOA65 at the LC-glass interface in the presence of BSA (Figure 2d), it is expected that the average tilt angle may change further so that signal amplification contributed by polymerized NOA65 aggregates can be detected and quantitated through dielectric spectroscopy. When E7 was doped with 4-wt.% NAO65 and irradiated with UV at 20 mW/cm^2 for 0, 60, 180 or 300 s in the absence of BSA, ε' remained unchanged irrespective of exposure times (Figure 6b). The slight decrease in ε' at each exposure time compared with the parallel component of the dielectric constant of E7 ($\varepsilon_\parallel$ = 19.5), which represents the state where the LC molecules were vertically aligned, may be partially attributed to the doped NOA65, which has a dielectric constant of 4.6 [30].

At a doping concentration of 2-wt.% NOA65, the NOA65/E7 composite prepared by exposure to UV at 20 mW/cm^2 for 300 s exhibited a decrease in ε' with increasing BSA concentrations, whereas ε' was kept constant prior to UV irradiation (exposure time 0 s) (Figure 7a). When NOA65 concentration was increased to 4 wt.%, a similar inverse correlation to that at 2-wt.% NOA65 between ε' and BSA concentration was observed (Figure 7b). At each BSA concentration, prolonged UV exposure (Figure 7b) as well as an increase in NOA65 (compare Figure 7a,b at 300 s) resulted in lower ε'. The value of ε' decreased from 18.1 to 11.2 with increasing BSA concentration in the range of 10^{-13} to 10^{-5} g/mL at a NOA65 concentration of 4 wt.% and UV exposure of 20 mW/cm^2 for 300 s. These results confirm our findings in optical measurements and provide strong evidence that photopolymerized NOA65 enhanced the optical and dielectric signals in LC-based biosensing. The calculated LOD for the NOA65/E7 composite at 2-wt.% and 4-wt.% NOA65 with UV exposure at 20 mW/cm^2 for 300 s was 2.9×10^{-10} and 8.8×10^{-11}-g/mL BSA, respectively. When E7 doped with 4-wt.% NOA65 and exposed to 20 mW/cm^2 for 300 s was used as the sensing medium, the LOD determined by optical measurements was 3.4 × 10^{-9} g/mL for BSA, which was an order of magnitude higher than that obtained by dielectric measurements (8.8 × 10^{-11}-g/mL BSA). In our previously reported dielectric protein assay based on dual-frequency LC (DFLC), ε' at 100 kHz in the high-frequency regime increased from 3.74 ± 0.02 to 5.10 ± 0.04 while that at 200 Hz in the low-frequency regime decreased from 9.83 ± 0.04 to 8.46 ± 0.05 when BSA concentration was increased from 10^{-7} to 10^{-2} g/mL [16]. Although the absolute value of ε' varied with the type of LCs and the frequency at which ε' was measured, it was observed that the LOD was lowered by an order of magnitude (from 10^{-6} to 10^{-7} g/mL for BSA) compared with that determined by optical texture observation when BSA was quantitatively analyzed by DFLC-based dielectric measurements [16], which was in agreement with the findings of this study. Results from this and our previous studies therefore imply that dielectric spectroscopic analysis offers more sensitive detection of biomolecules in comparison with the qualitative or semi-quantitative optical measurements.

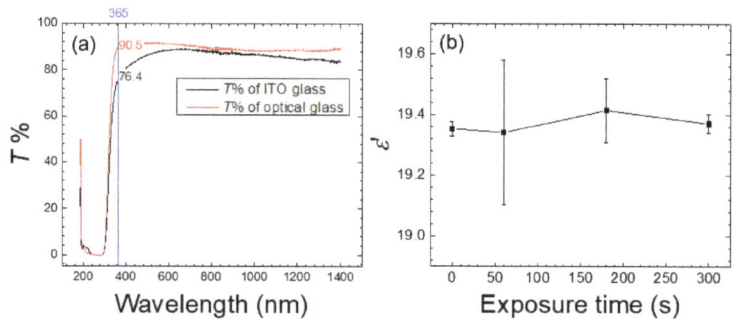

Figure 6. Transmission spectrometric analysis of glass substrates and dielectric measurement of E7 doped with 4-wt.% NOA65 in the absence of BSA. (**a**) Comparison of the optical transmission spectra of the optical and ITO-coated glass slides measured within a wavelength range of 200–1500 nm. (**b**) The real-part dielectric constant ε' determined at a UV irradiance of 20 mW/cm^2 with various exposure times. Error bars represent the standard deviation calculated from ε' of at least three independent experiments.

Calculated as $\varepsilon'_t/\varepsilon'_0$, where ε'_t and ε'_0 stand for ε' obtained at a UV exposure time t and 0 s, respectively, Figure 7c displays the reduced ε' against the BSA concentration deduced from Figure 7b. Expressing the dielectric signal using reduced ε' enabled the measured ε'_t under different experimental conditions to be normalized to a constant ε'_0, which represents the dielectric constant of the NOA65/E7 mixture prior to UV irradiation. As a result, the maximum value of reduced ε' is unity, corresponding to the unperturbed homeotropic state as in the absence of the analyte, whereas the minimum value is $\varepsilon_\perp/\varepsilon'_0 = 5.2/19.5 = 0.267$, reflecting a state where the anchoring effect of DMOAP diminished due to the accumulated analyte at the LC-glass interface. To demonstrate signal amplification by photopolymerized NOA65 more explicitly, the effective tilt angle θ in radians expressed by

$$\theta = \sin^{-1}\sqrt{\frac{\varepsilon_{\text{eff}} - \varepsilon_\perp}{\varepsilon_\parallel - \varepsilon_\perp}} = \sin^{-1}\sqrt{\frac{\varepsilon' - 5.2}{14.3}} \tag{3}$$

in accordance with Equation (2), was calculated for each ε' in Figure 7b. As shown in Figure 7d, θ decreased with increasing BSA concentration and UV exposure time, suggesting the nontrivial effect of NOA65 photopolymerization on the orientation of LC molecules.

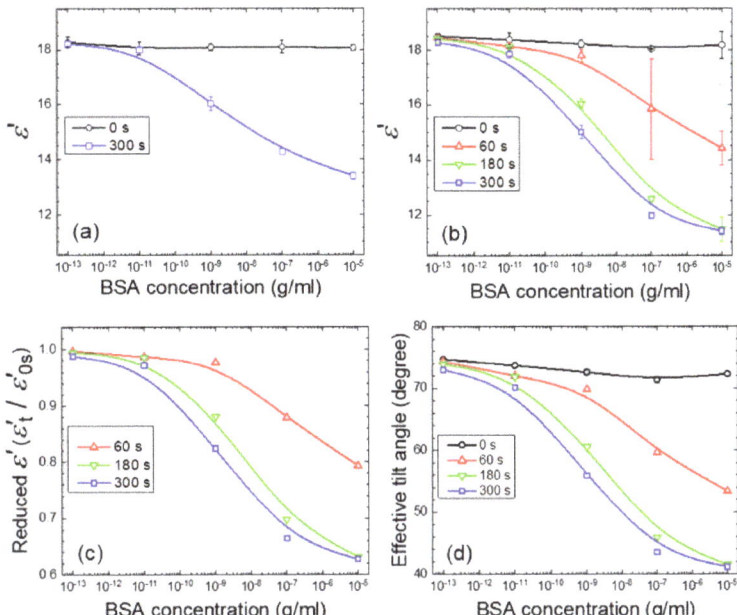

Figure 7. Dielectric spectroscopic analysis of NOA65/E7 cells exposed to 20-mW/cm² UV irradiation in the presence of BSA. (**a**) The real-part dielectric constant of E7 doped with 2-wt.% NOA65 and exposed to UV for 0 or 300 s. (**b**) The real-part dielectric constant, (**c**) the reduced ε', and (**d**) the effective tilt angle θ of E7 doped with 4-wt.% NOA65 and exposed to UV for 0, 60, 180, or 300 s as a function of the BSA concentration. The value of θ was calculated by Equation (3) based on the data retrieved from (**b**). Curves are based on spline fitting.

To mathematically describe the correlation between the BSA concentration c and the real-part dielectric constant ε', additional dielectric measurements were taken from supplementary samples with BSA concentrations at 10^{-12}, 10^{-10}, 10^{-8}, and 10^{-6} g/m and the results were combined with the data shown in Figure 7b for the exposure time of 300 s. Figure 8 shows the complete set of the nine experimental data points and two curves fitted to polynomials of the third order with the coefficient of determination $R^2 = 0.9830$ to cover the entire BSA concentration range of 10^{-13}–10^{-5} g/mL or $R^2 = 0.9997$ for the narrower range of 10^{-11}–10^{-7} g/mL. This permits one to obtain the BSA concentration (in g/mL) simply from the measured ε' value using the following equation:

$$\log(c) = B_0 + B_1 \varepsilon' + B_2 \varepsilon'^2 + B_3 \varepsilon'^3 \tag{4}$$

where the coefficients are displayed in Table 1. The polynomial curve for the wider range of BSA concentration (10^{-13}–10^{-5} g/mL) in Figure 8 can be considered as consisting of three segments, with higher slopes at the two extremes compared to the intermediate segment. In the low BSA concentration range of 10^{-13}–10^{-11} g/mL, the measured ε' values decreased from 18.2 to 17.9, but were still close to the $\varepsilon'_\parallel$ value of 19.5, suggesting that the vertical anchoring strength of DMOAP dominated the control of the LC tilt angles, and the disturbance in the homeotropic alignment of LCs caused by the analyte was relatively weak. When BSA concentration was increased to the range of 10^{-11}–10^{-7} g/mL, ε' decreased further from 17.9 to 12, which indicates that the amount of BSA reached a critical value to mask the alignment effect of DMOAP, and the decrease in dielectric signal was predominantly determined by and proportional to the amount of immobilized BSA. At high BSA concentrations of 10^{-7}–10^{-5} g/mL, the extent of decrease in ε' (from 12 to

11.4) was again diminished. This can be explained by the saturation of BSA adsorbed on the DMOAP-coated glass surface. Increasing the amount of BSA would not give rise to proportional decrease in ε' as observed in the 10^{-11}–10^{-7} g/mL concentration range, because excess BSA that can no longer adsorb to the DMOAP-coated glass surface was washed away during the sample preparation process. Therefore, by eliminating the ε' data in the BSA concentration ranges of 10^{-13}–10^{-11} and 10^{-7}–10^{-5} g/mL during curve fitting, the third-order polynomial fitting curve (red dashed curve, Figure 8) coincides with the linear regression line (green dashed line, Figure 8) between 10^{-11} and 10^{-7} g/mL BSA. The monotonic correlation thus revealed between ε' and BSA concentration supports the above explanation on the dominant effect of the amount of BSA on ε' in the 10^{-11}–10^{-7} g/mL concentration range.

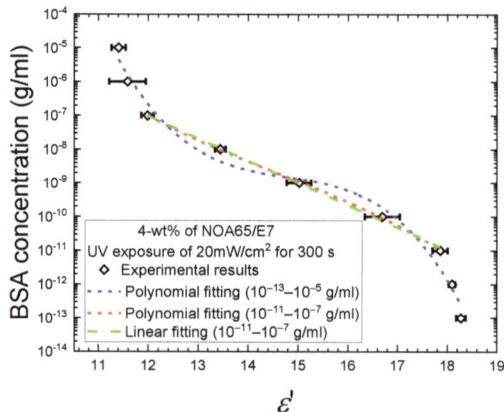

Figure 8. The BSA concentration as a function of the measured dielectric value, allowing interpolation of the concentration of the protein analyte in the dielectric permittivity range between 11.4 and 18.3. The blue and red dashed curves are third-order polynomial functions describing the relation of ε' to BSA concentration in the wider 10^{-13}–10^{-5} g/mL and the narrower 10^{-11}–10^{-7} g/mL range, respectively. The green dashed line represents the result of linear regression in the BSA concentration range of 10^{-11}–10^{-7} g/mL.

Table 1. Polynomial coefficients (in g/mL) in Equation (4) obtained through curve fitting of the experimental data as given in Figure 8.

c Range (g/mL)	B_0	B_1	B_2	B_3	R^2
10^{-13}–10^{-5}	238.07437 ± 69.99535	−49.61891 ± 14.72656	3.33539 ± 1.01777	−0.07501 ± 0.02311	0.98304
10^{-11}–10^{-7} (polynomial)	32.18531 ± 15.9708	−7.23333 ± 3.30443	0.45545 ± 0.22561	−0.0104 ± 0.00508	0.99965
10^{-11}–10^{-7} (linear)	0.93695 ± 0.25101	−0.66235 ± 0.01734	0	0	0.99795

To further simplify the mathematical expression [15], linear regression using the method of least squares was performed on the c–θ curve for the UV exposure time of 300 s in Figure 7d. To improve the accuracy of linear regression analysis, the ε' and θ values for additional BSA samples of 10^{-12}, 10^{-10}, 10^{-8}, and 10^{-6} g/mL were measured and calculated, respectively (data not shown). When expressed by the following linear correlation,

$$\theta = a \log(c) + b \qquad (5)$$

where a stands for the slope and b the vertical intercept of the linear function θ (c), a calibration curve was obtained from which the concentration of an unknown protein sample can be interpolated with its θ value. Note that Equation (5) is apparently invalid for $c = 0$ and, to be conservative, it is limited to the experimental range of 10^{-13} g/mL $\leq c \leq 10^{-5}$ g/mL. Such linear and inverse correlation between θ and c was also derived in our previous work on a capacitive biosensor based on a LC of high birefringence to describe the variation of θ over a BSA concentration range of 10^{-9} to 10^{-3} g/mL [15]. As presented in Table 2, the R-squared value for the regression analysis was 0.974 for the entire BSA concentration range of 10^{-13}–10^{-5} g/mL. Because in most biochemical assays the range of protein concentration in an analyte usually spans only two to three orders of magnitude, two narrower ranges of linearity, 10^{-11}–10^{-7} and 10^{-10}–10^{-8} g/mL BSA, were selected for comparison (Table 2). By reducing the concentration range for linear fitting, the R-squared value and thus the accuracy of protein quantitation can be increased.

Table 2. Linear regression parameters for the plot of effective tilt angle θ (in degrees) versus BSA concentration c (in g/mL). Regression analysis was performed on the θ values within three different BSA concentration ranges for comparison.

c (g/mL)	a (°)	b (°)	R^2
10^{-13}–10^{-5}	−4.97422	11.7882	0.97422
10^{-11}–10^{-7}	−6.43209	−1.74649	0.99863
10^{-10}–10^{-8}	−6.55982	−3.09482	0.99985

4. Conclusions

An optical and dielectric protein biosensor based on a LC–photopolymer composite was established in this study. Compared to our previously reported single-substrate detection, the sensing platform constructed with LC cells enabled the generation of a uniform electric field between two parallel conducting glass surfaces for dielectric spectroscopic measurements, which is a crucial advantage as biosensing in conjunction with dielectric spectroscopy led to improved sensitivity and LOD. Through optical texture observation and image analysis, it was demonstrated that by synthesizing a LC–photopolymer composite consisting of E7 impregnated with 4-wt.% NOA65, followed by UV irradiation at 20 mW/cm^2 for 300 s, significant signal amplification was achieved for the detection of BSA. The photopolymerized NOA65, but not the prepolymer prior to UV exposure, contributed to the enhanced optical signal, and UV irradiation had no effect on the brightness of the optical texture of pristine E7 in the absence of NOA65. By subjecting the LC–photopolymer composite to an externally applied electric field, dielectric spectroscopic analysis was performed to improve the sensitivity and LOD (88 pg/mL BSA, determined by dielectric measurements), offering a novel means of quantitative protein assay. Investigating the BSA concentration dependence of the real-part dielectric constant of the LC composite led to the calculation of the effective tilt angle, which significantly decreased only when NOA65 was photopolymerized by UV. These findings strongly support that photopolymerized NOA65 altered the LC orientation to enhance the transduced optical and dielectric signals of BSA. The optical and dielectric biosensing technique based on the NOA65/E7 composite is a label-free end-point assay, which can be easily adopted to a wide variety of biochemical and clinical assays such as immunoassays and enzyme assays that rely on biomolecular interactions on a solid substrate.

Author Contributions: Conceptualization, M.-J.L. and W.L.; Methodology, S.-C.Y.; Software, S.-C.Y.; Validation, H.S., M.-J.L. and W.L.; Formal Analysis, H.S. and S.-C.Y.; Investigation, S.-C.Y.; Resources, M.-J.L. and W.L.; Data Curation, H.S. and S.-C.Y.; Writing—Original Draft Preparation, H.S.; Writing—Review & Editing, M.-J.L. and W.L.; Visualization, H.S. and S.-C.Y.; Supervision, M.-J.L. and W.L.; Project Administration, M.-J.L. and W.L.; Funding Acquisition, M.-J.L. and W.L. All authors have read and agreed to the published version of the manuscript.

Funding: This research was funded by the Ministry of Science and Technology, Taiwan, under grant Nos. 107-2112-M-009-012-MY3 and 109-2320-B-309-001.

Institutional Review Board Statement: Not applicable.

Informed Consent Statement: Not applicable.

Data Availability Statement: The authors confirm that the data supporting the findings of this study are available within the article.

Acknowledgments: The authors thank P.-C. Wu for useful discussion during S.-C.Y.'s experimental process.

Conflicts of Interest: The authors declare no conflict of interest.

References

1. Gupta, V.K.; Skaife, J.J.; Dubrovsky, T.B.; Abbott, N.L. Optical amplification of ligand-receptor binding using liquid crystals. *Science* **1998**, *279*, 2077–2080. [CrossRef]
2. Brake, J.M.; Daschner, M.K.; Abbott, N.L. Formation and characterization of phospholipid monolayers spontaneously assembled at interfaces between aqueous phases and thermotropic liquid crystals. *Langmuir* **2005**, *21*, 2218–2228. [CrossRef] [PubMed]
3. Brake, J.M.; Daschner, M.K.; Luk, Y.Y.; Abbott, N.L. Biomolecular interactions at phospholipid-decorated surfaces of liquid crystals. *Science* **2003**, *302*, 2094–2097. [CrossRef] [PubMed]
4. Brake, J.M.; Abbott, N.L. An Experimental system for imaging the reversible adsorption of amphiphiles at aqueous−liquid crystal interfaces. *Langmuir* **2002**, *18*, 6101–6109. [CrossRef]
5. Brake, J.M.; Mezera, A.D.; Abbott, N.L. Active control of the anchoring of 4′-pentyl-4-cyanobiphenyl (5CB) at an aqueous−liquid crystal interface by using a redox-active ferrocenyl surfactant. *Langmuir* **2003**, *19*, 8629–8637. [CrossRef]
6. Khan, M.; Khan, A.R.; Shin, J.H.; Park, S.Y. A liquid-crystal-based DNA biosensor for pathogen detection. *Sci. Rep.* **2016**, *6*, 22676. [CrossRef]
7. Chen, C.-H.; Yang, K.-L. A liquid crystal biosensor for detecting organophosphates through the localized pH changes induced by their hydrolytic products. *Sen. Actuators B Chem.* **2013**, *181*, 368–374. [CrossRef]
8. Hu, Q.-Z.; Jang, C.-H. A simple strategy to monitor lipase activity using liquid crystal-based sensors. *Talanta* **2012**, *99*, 36–39. [CrossRef] [PubMed]
9. Popova, P.; Honakerb, L.W.; Kooijmanc, E.E.; Manna, E.K.; Jáklib, A.I. A liquid crystal biosensor for specific detection of antigens. *Sens. Bio-Sens. Res.* **2016**, *8*, 31–35. [CrossRef]
10. Lee, M.-J.; Lee, W. Liquid crystal-based capacitive, electro-optical and dielectric biosensors for protein quantitation. *Liq. Cryst.* **2020**, *47*, 1145–1153. [CrossRef]
11. Hsiao, Y.-C.; Sung, Y.-C.; Lee, M.-J.; Lee, W. Highly sensitive color-indicating and quantitative biosensor based on cholesteric liquid crystal. *Biomed. Opt. Express* **2015**, *6*, 5033–5038. [CrossRef]
12. Lee, M.-J.; Chang, C.-H.; Lee, W. Label-free protein sensing by employing blue phase liquid crystal. *Biomed. Opt. Express* **2017**, *8*, 1712–1720. [CrossRef] [PubMed]
13. Chiang, Y.-L.; Lee, M.-J.; Lee, W. Enhancing detection sensitivity in quantitative protein detection based on dye-doped liquid crystals. *Dyes Pigment.* **2018**, *157*, 117–122. [CrossRef]
14. Wu, P.-C.; Karn, A.; Lee, M.-J.; Lee, W.; Chen, C.-Y. Dye-liquid-crystal-based biosensing for quantitative protein assay. *Dyes Pigment.* **2018**, *150*, 73–78. [CrossRef]
15. Lin, C.H.; Lee, M.J.; Lee, W. Bovine serum albumin detection and quantitation based on capacitance measurements of liquid crystals. *Appl. Phys. Lett.* **2016**, *109*, 093703. [CrossRef]
16. Lin, C.-M.; Wu, P.-C.; Lee, M.-J.; Lee, W. Label-free protein quantitation by dielectric spectroscopy of dual-frequency liquid crystal. *Sens. Actuators B Chem.* **2019**, *282*, 158–163. [CrossRef]
17. Li, X.; Li, G.; Yang, M.; Chen, L.-C.; Xiong, X.-L. Gold nanoparticle based signal enhancement liquid crystal biosensors for tyrosine assays. *Sens. Actuators B Chem.* **2015**, *215*, 152–158. [CrossRef]
18. Nandi, R.; Loitongbam, L.; De, J.; Jain, V.; Pal, S.K. Gold nanoparticle-mediated signal amplification of liquid crystal biosensors for dopamine. *Analyst* **2019**, *144*, 1110–1114. [CrossRef]
19. Wang, Y.; Wang, B.; Xiong, X.; Deng, S. Gold nanoparticle-based signal enhancement of an aptasensor for ractopamine using liquid crystal based optical imaging. *Microchim. Acta* **2019**, *186*, 697. [CrossRef]
20. Su, H.-W.; Lee, M.-J.; Lee, W. Surface modification of alignment layer by ultraviolet irradiation to dramatically improve the detection limit of liquid-crystal-based immunoassay for the cancer biomarker CA125. *J. Biomed. Opt.* **2015**, *20*, 57004. [CrossRef]
21. Su, H.-W.; Lee, Y.-H.; Lee, M.-J.; Hsu, Y.-C.; Lee, W. Label-free immunodetection of the cancer biomarker CA125 using high-Δn liquid crystals. *J. Biomed. Opt.* **2014**, *19*, 077006. [CrossRef]
22. Sun, S.-H.; Lee, M.-J.; Lee, Y.-H.; Lee, W.; Song, X.; Chen, C.-Y. Immunoassays for the cancer biomarker CA125 based on a large-birefringence nematic liquid-crystal mixture. *Biomed. Opt. Express* **2015**, *6*, 245–256. [CrossRef]
23. Hsu, W.-L.; Lee, M.-J.; Lee, W. Electric-field-assisted signal amplification for label-free liquid-crystal-based detection of biomolecules. *Biomed. Opt. Express* **2019**, *10*, 4987–4998. [CrossRef]
24. Lee, M.-J.; Duan, F.-F.; Wu, P.-C.; Lee, W. Liquid crystal-photopolymer composite films for label-free single-substrate protein quantitation and immunoassay. *Biomed. Opt. Express* **2020**, *11*, 4915–4927. [CrossRef]

25. Hsu, C.J.; Chen, B.L.; Huang, C.Y. Controlling liquid crystal pretilt angle with photocurable prepolymer and vertically aligned substrate. *Opt. Express* **2016**, *24*, 1463–1471. [CrossRef] [PubMed]
26. Hsu, C.J.; Cui, Z.Y.; Chiu, C.-C.; Hsiao, F.-L.; Huang, C.Y. Self-assembled polymer gravel array in prepolymer-doped nematic liquid crystals. *Opt. Mater. Express* **2017**, *7*, 4374–4385. [CrossRef]
27. Yang, K.H. Measurements of empty cell gap for liquid-crystal displays using interferometric methods. *J. Appl. Phys.* **1988**, *64*, 4780–4781. [CrossRef]
28. Lu, J.-P.; Huang, W.-K.; Chen, F.-C. Self-positioning microlens arrays prepared using ink-jet printing. *Opt. Eng.* **2009**, *48*, 073606. [CrossRef]
29. Shrivastava, A.; Gupta, V. Methods for the determination of limit of detection and limit of quantitation of the analytical methods. *Chron. Young Sci.* **2011**, *2*, 21–25. [CrossRef]
30. Jisha, C.P.; Hsu, K.-C.; Lin, Y.; Lin, J.-H.; Chuang, K.-P.; Tai, C.-Y.; Lee, R.-K. Phase separation and pattern instability of laser-induced polymerization in liquid-crystal-monomer mixtures. *Opt. Mater. Express* **2011**, *1*, 1494–1501. [CrossRef]

Communication

Highly Sensitive Detection of CA 125 Protein with the Use of an n-Type Nanowire Biosensor

Kristina A. Malsagova [1,*], Tatyana O. Pleshakova [1], Rafael A. Galiullin [1], Andrey F. Kozlov [1], Ivan D. Shumov [1], Vladimir P. Popov [2], Fedor V. Tikhonenko [2], Alexander V. Glukhov [3], Vadim S. Ziborov [1,4], Oleg F. Petrov [4], Vladimir E. Fortov [4], Alexander I. Archakov [1] and Yuri D. Ivanov [1]

1. Laboratory of nanotechnology, Institute of Biomedical Chemistry, 119121 Moscow, Russia; t.pleshakova1@gmail.com (T.O.P.); rafael.anvarovich@gmail.com (R.A.G.); afkozlow@mail.ru (A.F.K.); shum230988@mail.ru (I.D.S.); ziborov.vs@yandex.ru (V.S.Z.); alexander.archakov@ibmc.msk.ru (A.I.A.); yurii.ivanov.nata@gmail.com (Y.D.I.)
2. Rzhanov Institute of Semiconductor Physics, Siberian Branch of Russian Academy of Sciences, 630090 Novosibirsk, Russia; popov@isp.nsc.ru (V.P.P.); ifp@isp.nsc.ru (F.V.T.)
3. JSC Novosibirsk Plant of Semiconductor Devices with OKB, 630082 Novosibirsk, Russia; gluhov@nzpp.ru
4. Joint Institute for High Temperatures of Russian Academy of Sciences, 125412 Moscow, Russia; ofpetrov@ihed.ras.ru (O.F.P.); fortov@ihed.ras.ru (V.E.F.)
* Correspondence: kristina.malsagova86@gmail.com; Tel.: +7-499-246-3761

Received: 20 November 2020; Accepted: 17 December 2020; Published: 18 December 2020

Abstract: The detection of CA 125 protein in a solution using a silicon-on-insulator (SOI)-nanowire biosensor with n-type chip has been experimentally demonstrated. The surface of nanowires was modified by covalent immobilization of antibodies against CA 125 in order to provide the biospecificity of the target protein detection. We have demonstrated that the biosensor signal, which results from the biospecific interaction between CA 125 and the covalently immobilized antibodies, increases with the increase in the protein concentration. At that, the minimum concentration, at which the target protein was detectable with the SOI-nanowire biosensor, amounted to 1.5×10^{-16} M.

Keywords: ovarian cancer; nanowire biosensor; nanowire; silicon-on-insulator; CA 125; antibodies

1. Introduction

The effective treatment of most pathologies, including cancer, depends on the early revelation of a pathological process [1]. The identification of target biomarkers, associated with the early-stage (asymptomatic) development of a disease, is a starting point for choosing an appropriate and effective treatment. Most of the protein markers are present in the blood at low (<10^{-13} M) or ultra-low (<10^{-15} M) concentrations. The blood concentration of cancer biomarkers at the early stage of an oncological disease is at the level of 10^{-15} M (that is, at the femtomolar level), as was emphasized by Rissin et al. [2]. The early revelation of cancer in human, accordingly, requires the development of novel methods, which allow for one to detect cancer biomarkers with, at least, femtomolar concentration sensitivity. The application possibilities of standard immunohistochemical, radioimmunoassay (RIA)-based, enzyme-linked immunosorbent assay (ELISA)-based methods, etc. (which are commonly employed in clinical diagnostics for the detection of protein markers), are limited due to: (1) their low (10^{-14} M to 10^{-7} M) concentration sensitivity and (2) the need to use enzyme and fluorescent labels.

The use of nanowire biosensors gives new opportunities for biomedical research, as well as for clinical practice in the future. One of the key advantages of this type of biosensors consists in the possibility of direct label-free detection of a target protein in real-time with high (<10^{-15} M) concentration sensitivity [3]. The operation of a nanowire biosensor is based on the registration of a modulation of the electric current through the nanowire sensor elements upon adsorption of target protein molecules onto the surface of the sensor elements. The surface-adsorbed molecules act as a virtual gate, and the nanowire structure itself with ohmic contacts on its ends acts as a nanoscale field-effect transistor (FET) [4]. The high sensitivity of the nanowire sensor element is determined by its high surface-to-volume ratio [5]. The theoretical detection limit, which is attainable with a nanowire biosensor, can reach the level of a single molecule per sensor element [6]. In this way, F. Patolsky et al. [7] reported that the use of nanowire biosensors allows for the detection of viruses with the sensitivity at the single-particle level. Regarding biological macromolecules, the femtomolar detection limit was experimentally attained for DNA [8,9]; for proteins, an even lower (subfemtomolar) detection limit was attained [3].

The surface of nanowire sensor elements is functionalized with biospecific probe molecules (molecular probes) in order to provide the biospecificity of the detection of target protein markers of diseases. The formation of an array, containing multiple nanowires on a single chip, with subsequent functionalization of the nanowires with molecular probes against various types of target biomolecules represents another important advantage of the nanowire biosensors, since this allows for one to conduct multiplexed detection of target proteins in one sample. Thus, nanowire biosensors combine the following advantages: (1) highly sensitive label-free detection of target proteins and the (2) rapid simultaneous express analysis of a wide range of target proteins.

CA 125 protein represents a marker, which is associated with the development of malignant tumors (ovarian cancer, uterine cancer, endometrial cancer, breast cancer, etc.), benign tumors (endometriosis, pleurisy, etc.), and inflammatory diseases [10]. The discovery of this antigen has become an important step on the way to the development of a biochemical approach to the non-invasive diagnosis and monitoring of ovarian cancer. The use of CA 125 as a marker of ovarian cancer was suggested in 1983 and, since this time, it has been considered as the benchmark for monitoring ovarian cancer patients [11]. CA 125 represents a glycoprotein epitope of a mucin with high molecular weight [10]. One of the main causes of errors, which occurs during biomarker detection, is their biological variability [12]. The use of CA 125 is limited by the low sensitivity of the marker (<50%) for the initial stage of the disease and its poor specificity, especially in young women. However, the results of twenty-three randomized large-scale screening research studies on 250,000 women suggest the benefits of screening that is based on CA 125 evaluation, in order to early detect ovarian cancer in menopause-aged women, as well as in women with familial clustering of ovarian cancer [13].

Herein, antibodies against CA 125 have been employed as molecular probes for the functionalization of an n-type nanowire biosensor chip. This chip was fabricated on the basis of silicon-on-insulator (SOI) structure employing complementary metal–oxide–semiconductor (CMOS)-compatible technology. In contrast to our previous studies [14–18], before the surface functionalization, the sensor chips were treated with glow discharge plasma instead of ozone treatment. The antibody-functionalized chips have been used for the highly sensitive detection of high molecular weight glycoprotein CA 125—a protein marker of ovarian cancer—in purified buffer solution. The experimentally attained concentration detection limit of CA 125 was $\sim 10^{-16}$ M. Because the early diagnosis of oncological pathologies requires the use of highly sensitive detection methods, which allows for attaining a 10^{-15} M concentration detection limit [2], our nanowire biosensor represents an attractive tool for the rapid express analysis of protein markers of oncological diseases.

2. Materials and Methods

2.1. Chemicals

3,3′-dithiobis (sulfosuccinimidyl propionate) (DTSSP cross-linker) was purchased from Pierce (Waltham, MA, USA). Potassium phosphate monobasic (KH_2PO_4) and 3-aminopropyltriethoxysilane (APTES) were purchased from Sigma–Aldrich (St. Louis, MO, USA). Methanol (CH_3OH) was purchased from Sigma (St. Louis, MO, USA). Hydrofluoric acid (HF) was purchased from Reakhim (Moscow, Russia). Deionized water was obtained while using a Milli-Q system (Millipore, Molsheim, France).

2.2. Proteins

Monoclonal antibodies against CA125 (clone 13F4, isotype IgG1) were purchased from USBio (Salem, MA, USA). The recombinant CA125 protein (molecular weight 110 kDa; 10^{-6} M stock solution in potassium phosphate buffer) was purchased from R&D Systems (Minneapolis, MN, USA).

Antibodies against Bcl-2 protein were purchased from Biorbyt, Ltd. (Cambridge, UK).

2.3. Fabrication of Nanowire Sensors

The fabrication and characteristics of the SOI-nanowire sensor chips (SOI-NW chips) are described in detail elsewhere [15,16]. The process of silicon-on-insulator (SOI) nanowire chips fabrication, as schematically shown in Figure 1, comprised of the following steps: the production of initial SOI structures with a cut-off Si layer thickness of 500–600 nm while using hydrogen exfoliation technology; thinning of the SOI layer to nanometer dimensions by a sequential cycle of operations-thermal oxidation; removal of sacrificial oxide in HF solution; lateral structuring of the SOI layers using optical or electron lithography to form nanowire structures with contact areas; the formation of ohmic contacts to nanometer thick SOI layer by thickening the SOI in the contact areas by a poly-Si layer deposition and subsequent doping; lateral structuring of SOI layers while using electronic lithography and gas-plasma chemical etching, which allows for one to form a nanometer-size active element; metallization and contact wiring; and, finally, crystal cutting.

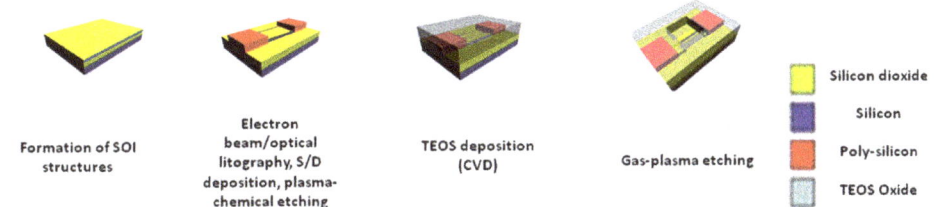

Figure 1. Schematic representation of silicon-on-insulator-NW (SOI-NW) sensor fabrication.

In our biosensor, SOI-NW chips with n-type conductance were employed. The thickness of the cut-off silicon layer was 32 nm and the buried oxide (BOX) thickness was 300 nm. The width of the nanowire sensor elements was $w = 3$ μm, while their thickness and length were $t = 32$ nm and $l = 10$ μm, respectively. The number of nanowires on the crystal was 12. Figure 2 displays the typical SEM image of a single nanowire sensor element.

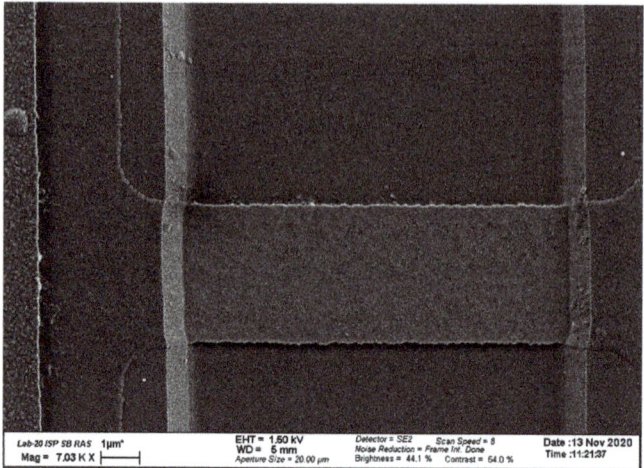

Figure 2. Typical SEM image of a single nanowire sensor element.

2.4. Modification of the Surface of the SOI-Nanowire Chip

The surface of the SOI-NW chips was first treated chemically, in order to remove the organic contaminants and the natural oxide from the sensor surface, with isopropanol, HF, and CH_3OH similarly to the procedure that was described in our previous papers [14–18]. After that, the chips were treated with glow discharge plasma in a homemade apparatus that was developed in JIHT RAS, in order to form OH groups on the sensor surface, and then the chip was treated in APTES vapors, according to [17,18].

2.5. Covalent Immobilization of Molecular Probes

The molecular probes (antibodies against CA 125 and against Bcl-2) were covalently immobilized onto the chemically modified surface of the nanowires with the use of the DTSSP crosslinker. For this purpose, 3-nL microdrops of 0.8 μM solutions of antibodies in potassium phosphate (KP) buffer (50 mm, pH 7.4) were precisely dispensed onto the surface of individual DTSSP-activated nanowires with a Piezorray micro-arraying system (PerkinElmer, Inc., Waltham, MA, USA). The solutions were incubated on the surface of the nanowires for 30 min. at 15 °C and 80% humidity. After that, the surface of the chip was washed with deionized water for 30 min.

2.6. Preparation of CA 125 Solutions in Buffer

CA 125 solutions with concentrations that ranged from 10^{-18} M to 10^{-15} M were prepared from the initial stock solution of the protein (1 μM in 50 mM KP, pH 7.4) by sequential tenfold dilution with 1 mM KP buffer (pH 7.4). On each dilution step, the protein solution was incubated in a shaker at 10 °C for 30 min. The so-prepared protein solutions were then immediately used in the biosensor measurements.

2.7. Electrical Measurements

The nanowire biosensor setup is described in detail in [19]. The electrical measurements were performed with a Keithley Model 6487 picoampermeter (Keithley, Solon, USA). During the measurements, the substrate of the SOI structures was used as the control electrode (transistor gate). In the course of the experiments, the dependence of source-drain current on gate voltage $I_{ds}(V_g)$ at V_g from 0 to 100 V and $V_{ds} = 0.15$ V was obtained for the SOI-NW chip. In order to detect the target protein, 150 μL of CA 125 solution in 1 mM KP buffer (pH 7.4) was added into the measuring cell containing 300 μL of buffer solution. The time dependencies of the current $I_{ds}(t)$ were recorded at $V_g = +50$ V and V_{ds} = 0.15 V. We used an additional Pt electrode, which was immersed into the solution in the measuring cell (similar to [3]), in order to increase the time stability of the nanowire sensors with a thin nanoconductor.

The biosensor included a 500-µL measuring cell, and the sensor chip with an array of nanowires served as the cell bottom. The diameter of the sensitive area was 2 mm. The solution in the cell was stirred at 3000 rpm.

3. Results

The detection of CA 125 was carried out in the measuring cell of the nanowire biosensor, while using a SOI-NW chip bearing an array of twelve 3-µm-thick n-type nanowires. The nanowires were functionalized by covalent immobilization of antibodies against CA 125 onto their surface in order to provide biospecificity of the CA 125 detection (as described in Materials and Methods). In order to account for the non-specific signal, a pair of nanowires of the same thickness (3 µm) was functionalized with antibodies against Bcl-2. The signal from these sensors was taken into account in order to calculate the resulting differential signal.

The sensograms were recorded before and after the addition of the CA 125 solutions in 1 mM KP buffer (pH 7.4) with the protein concentration ranging from 10^{-18} M to 10^{-14} M into the measuring cell of the biosensor.

Figure 3 displays typical sensograms that were recorded before and after the addition of CA 125 solutions into the measuring cell. The sensogram curves indicate that the addition of CA 125 solution leads to an increase in the signal from the nanowires with immobilized antibodies, owing to the binding of the target analyte molecules to their surface (Figure 3). Figure 3 shows that, in the case of using the SOI-NW chip with immobilized antibodies, the biosensor signal was clearly distinguishable until reaching the target protein concentration of 1.5×10^{-16} M.

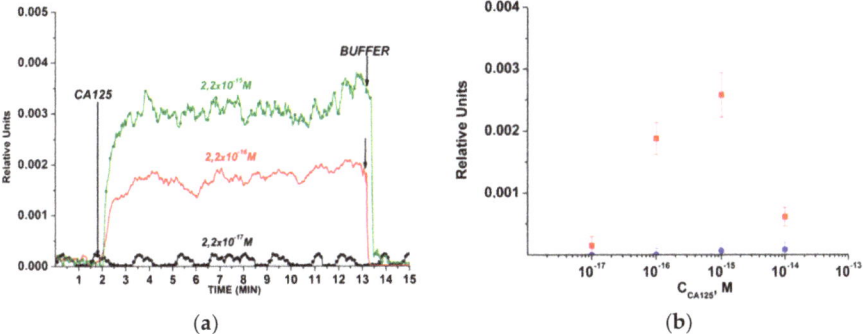

Figure 3. The results obtained upon the detection of CA125 protein in buffer solution while using an n-type SOI-NW chip with covalently immobilized antibodies: (**a**) typical sensograms obtained upon analysis of solutions with various concentrations of the target protein; (**b**) dependencies of the level of the biosensor signal on the concentration of CA 125 in buffer solution. The number of technical replicates was n = 3. Circles (●) and squares (■) indicate the average value of the signal level before and after the addition of the protein solution, respectively. The experimental conditions: 1 mM potassium phosphate (KP) buffer, pH 7.4, V_g = +50 V; V_{ds} = 0.15 V. The total volume of the solution in the cell was 450 µL. Arrows indicate the addition of the CA 125 solution (with concentrations from 2×10^{-18} to 2×10^{-14} M, as indicated in the Figure) and the wash with pure KP buffer.

The control experiments were carried out in order to determine the non-specific influence of the protein-free buffer on the biosensor signal. In the control experiments, upon the addition of the analyte-free working buffer into the measuring cell, either no response from the nanowires was observed or this response was no greater than 1 to 2% of the baseline signal level.

Moreover, a decrease in the response signal from the nanowires with decreasing the target protein concentration from 10^{-15} M to 10^{-17} M was observed.

These results allow for us to make a conclusion regarding the presence of the biospecific interaction between the molecular probes, immobilized on the SOI-NW chip surface, and the target protein molecules that are captured from the analyzed solution onto this surface.

It should be emphasized that the substitution of CA 125 solution with a protein-free buffer solution led to a decrease in the signal from the nanowires; in other words, it caused the dissociation of the CA 125/(antibodies against CA 125) complexes due to the shift in the biospecific interaction's equilibrium. This fact indicates the possibility of repeated use of the SOI-NW sensor chip for the detection of CA 125.

Upon increasing the target protein concentration to 2.2×10^{-14} M and higher values, no difference between the signal, which is received from the control nanowires, and that from working nanowires, was observed. This fact can be explained by the high degree of non-specific binding of target molecules to the surface of the control nanowire. A large number of molecules can lead to the oligomerization of the target protein and, consequently, to a change in the physicochemical parameters of the interaction of the target analyte molecules with the sensor surface—for instance, to a change in the efficiency of the protein adsorption onto the surface of the control nanowire.

The results obtained herein indicate that the immobilized molecular probes retain their affinity properties, and this allows for the biospecific capturing of the target protein onto the sensor surface. In our experiments, the lowest concentration of the target CA 125 protein, which was detectable with the antibody-functionalized SOI-NW chip, was 2.2×10^{-16} M.

In our present study, we have demonstrated the possibility of the nanowire biosensor-based detection of CA 125 oncomarker, employing purified solutions of a commercial CA 125 preparation in buffer and antibody-functionalized nanowire sensor chips, attaining a 10^{-16} M concentration detection limit. It should be emphasized that, in the case of viral infections (such as HCV infection [17]), their nucleic acid markers can be detected while using a polymerase chain reaction (PCR)-based assay. Moreover, nucleic acid molecules bear a large amount of negative charge—in contrast to the case with the majority of proteins—and, hence, represent objects that are much more easily detectable with a nanowire biosensor. In contrast, the diagnosis of ovarian cancer (as well as other oncological diseases) in human requires the detection of protein markers, and this is the approach that we develop in our present study. It is known that the early diagnosis of oncological pathologies requires the use of highly sensitive detection methods, which allow for one to attain a 10^{-15} M concentration detection limit [2]. In this respect, nanowire biosensor allows for one to overcome the 10^{-15} M sensitivity threshold, thus representing a quite attractive tool for the rapid detection of protein markers of oncological diseases. Moreover, the 10^{-16} M detection limit, attained with the use of a nanowire biosensor in our present study, is not an ultimate point, and it can be further shifted down. One of the ways to lower the detection limit is the use of a scheme involving a microwave generator, as was reported in one of our previous papers [17]. In addition, decreasing the width of the nanowire sensor elements, providing higher surface-to-volume ratio [5], is another way for further increasing the sensitivity of nanowire biosensors. In principle, decreasing the nanowire width can allow for the single-charge sensitivity of the biosensor [5], which means the possibility to perform nanowire-based detection of charged biomolecules with single-molecule sensitivity, and this is what we will aim for in future research.

Moreover, our nanowire sensor chips are fabricated while using a CMOS-compatible technology, and this is another advantage of the biosensor proposed herein, as it allows for the transition to large-scale production, providing low cost of the sensor chips—which is required for the clinical screening applications of the approach being developed.

4. Conclusions

Herein, the highly sensitive detection of cancer-associated protein marker CA 125 in buffer solution (at pH 7.4) with a nanowire biosensor has been experimentally demonstrated. Silicon-on-insulator

(SOI) structures, which were fabricated using top-down technology, were used as sensor chips. For the functionalization of the sensor surface, antibodies against CA 125 were used as biospecific molecular probes. The use of antibody-functionalized SOI-NW chips has allowed us to experimentally attain the concentration limit of CA 125 detection at the level of 2.2×10^{-16} M.

The results obtained herein indicate that the nanowire biosensor represents a prototype of a medical diagnostic device, which can be employed for the revelation of cancer. Moreover, because our nanowire biosensor includes a sensor chip bearing an array of 12 nanowires, its application will allow for one to perform the simultaneous selective early diagnosis of a number of common and socially significant diseases in one test, which seems to be promising for screening applications in medical diagnostics.

Author Contributions: Conceptualization, Y.D.I., V.P.P. and A.I.A.; methodology, T.O.P.; software, R.A.G. and V.S.Z.; validation, T.O.P. and V.E.F.; formal analysis, K.A.M., V.S.Z. and A.V.G.; investigation, K.A.M., A.F.K., R.A.G., F.V.T. and V.S.Z.; resources, V.P.P., A.V.G., V.E.F. and O.F.P.; visualization, K.A.M., I.D.S. and A.F.K.; data curation, F.V.T. and A.F.K.; writing—original draft preparation, K.A.M., T.O.P. and I.D.S.; writing—review and editing, Y.D.I.; project administration, Y.D.I.; supervision, A.I.A.; funding acquisition, A.I.A. All authors have read and agreed to the published version of the manuscript.

Funding: This work was financed by the Ministry of Science and Higher Education of the Russian Federation within the framework of state support for the creation and development of World-Class Research Centers "Digital biodesign and personalized healthcare" No. 75-15-2020-913.

Acknowledgments: The biosensor measurements were performed employing a nanowire detector, which pertains to "Avogadro" large-scale research facilities.

Conflicts of Interest: The authors declare no conflict of interest.

References

1. Johari-Ahar, M.; Rashidi, M.; Barar, J.; Aghaie, M.; Mohammadnejad, D.; Ramazani, A.; Karami, P.; Coukos, G.; Omidi, Y.; Agaie, M. An ultra-sensitive impedimetric immunosensor for detection of the serum oncomarker CA-125 in ovarian cancer patients. *Nanoscale* **2015**, *7*, 3768–3779. [CrossRef]
2. Rissin, D.M.; Kan, C.W.; Campbell, T.G.; Howes, S.C.; Fournier, D.R.; Song, L.; Piech, T.; Patel, P.P.; Chang, L.; Rivnak, A.J.; et al. Single-molecule enzyme-linked immunosorbent assay detects serum proteins at subfemtomolar concentrations. *Nat. Biotechnol.* **2010**, *28*, 595–599. [CrossRef]
3. Tian, R.; Regonda, S.; Gao, J.; Liu, Y.; Hu, W. Ultrasensitive protein detection using lithographically defined Si multi-nanowire field effect transistors. *Lab Chip* **2011**, *11*, 1952. [CrossRef]
4. Patolsky, F.; Zheng, G.; Lieber, C.M. Fabrication of silicon nanowire devices for ultrasensitive, label-free, real-time detection of biological and chemical species. *Nat. Protoc.* **2006**, *1*, 1711–1724. [CrossRef]
5. Elfström, N.; Juhasz, R.; Sychugov, I.; Engfeldt, T.; Karlström, A.E.; Linnros, J. Surface Charge Sensitivity of Silicon Nanowires: Size Dependence. *Nano Lett.* **2007**, *7*, 2608–2612. [CrossRef]
6. Hahm, J.-I.; Lieber, C.M. Direct Ultrasensitive Electrical Detection of DNA and DNA Sequence Variations Using Nanowire Nanosensors. *Nano Lett.* **2004**, *4*, 51–54. [CrossRef]
7. Patolsky, F.; Zheng, G.; Hayden, O.; Lakadamyali, M.; Zhuang, X.; Lieber, C.M. Electrical detection of single viruses. *Proc. Natl. Acad. Sci. USA* **2004**, *101*, 14017–14022. [CrossRef]
8. Lin, C.-H.; Hung, C.-H.; Hsiao, C.-Y.; Lin, H.-C.; Ko, F.-H.; Yang, Y.-S. Poly-silicon nanowire field-effect transistor for ultrasensitive and label-free detection of pathogenic avian influenza DNA. *Biosens. Bioelectron.* **2009**, *24*, 3019–3024. [CrossRef] [PubMed]
9. Wenga, G.; Jacques, E.; Salaün, A.-C.; Rogel, R.; Pichon, L.; Geneste, F. Bottom-gate and Step-gate Polysilicon Nanowires Field Effect Transistors for Ultrasensitive Label-free Biosensing Application. *Procedia Eng.* **2012**, *47*, 414–417. [CrossRef]
10. Kushlinskii, N.E.; Krasil'nikov, M.A. *Biological Tumor Markers: Basic and Clinical Research*; RAMS Publishing House: Moscow, Russia, 2017; p. 632.
11. Bast, R.C.; Feeney, M.; Lazarus, H.; Nadler, L.M.; Colvin, R.B.; Knapp, R.C. Reactivity of a monoclonal antibody with human ovarian carcinoma. *J. Clin. Investig.* **1981**, *68*, 1331–1337. [CrossRef] [PubMed]
12. Bast, R.C., Jr. Status of Tumor Markers in Ovarian Cancer Screening. *J. Clin. Oncol.* **2003**, *21* (Suppl. 10), 200s–205s. [CrossRef]

13. Fraser, C.G. Inherent biological variation and reference values. *Clin. Chem. Lab. Med.* **2004**, *42*, 758–764. [CrossRef] [PubMed]
14. Ivanov, Y.D.; Pleshakova, T.O.; Kozlov, A.F.; Malsagova, K.A.; Krohin, N.V.; Shumyantseva, V.V.; Shumov, I.D.; Tyschenko, I.; Naumova, O.V.; Fomin, B.I.; et al. SOI nanowire for the high-sensitive detection of HBsAg and α-fetoprotein. *Lab Chip* **2012**, *12*, 5104–5111. [CrossRef] [PubMed]
15. Malsagova, K.A.; Ivanov, Y.D.; Pleshakova, T.O.; Kaysheva, A.L.; Shumov, I.D.; Kozlov, A.F.; Archakov, A.I.; Popov, V.P.; Fomin, B.I.; Latyshev, A.V. A SOI-nanowire biosensor for the multiple detection of D-NFATc1 protein in the serum. *Anal. Methods* **2015**, *7*, 8078–8085. [CrossRef]
16. Ivanov, Y.D.; Pleshakova, T.O.; Kozlov, A.F.; Mal'Sagova, K.A.; Krokhin, N.V.; Kaisheva, A.L.; Shumov, I.D.; Tyschenko, I.; Naumova, O.; Fomin, B.I.; et al. SOI nanowire transistor for detection of D-NFATc1 molecules. *Optoelectron. Instrum. Data Process.* **2013**, *49*, 520–525. [CrossRef]
17. Malsagova, K.A.; Pleshakova, T.O.; Galiullin, R.A.; Kaysheva, A.L.; Shumov, I.D.; Ilnitskii, M.A.; Tyschenko, I.; Glukhov, A.V.; Archakov, A.I.; Ivanov, Y. Ultrasensitive nanowire-based detection of HCVcoreAg in the serum using a microwave generator. *Anal. Methods* **2018**, *10*, 2740–2749. [CrossRef]
18. Ivanov, Y.D.; Pleshakova, T.; Malsagova, K.; Kozlov, A.; Kaysheva, A.L.; Shumov, I.; Galiullin, R.; Kurbatov, L.; Popov, V.; Naumova, O.; et al. Detection of marker miRNAs in plasma using SOI-NW biosensor. *Sens. Actuators B Chem.* **2018**, *261*, 566–571. [CrossRef]
19. Malsagova, K.; Pleshakova, T.O.; Galiullin, R.A.; Shumov, I.D.; Kozlov, A.F.; Romanova, T.S.; Tyschenko, I.; Glukhov, A.V.; Konev, V.A.; Archakov, A.I.; et al. Nanowire Aptamer-Sensitized Biosensor Chips with Gas Plasma-Treated Surface for the Detection of Hepatitis C Virus Core Antigen. *Coatings* **2020**, *10*, 753. [CrossRef]

Publisher's Note: MDPI stays neutral with regard to jurisdictional claims in published maps and institutional affiliations.

© 2020 by the authors. Licensee MDPI, Basel, Switzerland. This article is an open access article distributed under the terms and conditions of the Creative Commons Attribution (CC BY) license (http://creativecommons.org/licenses/by/4.0/).

Review

Printed Electrochemical Biosensors: Opportunities and Metrological Challenges

Emilio Sardini [1], Mauro Serpelloni [1] and Sarah Tonello [2,*]

[1] Department of Information Engineering, University of Brescia, Via Branze 38, 25123 Brescia, Italy; emilio.sardini@unibs.it (E.S.); mauro.serpelloni@unibs.it (M.S.)
[2] Department of Information Engineering, University of Padova, Via Gradenigo 6, 35131 Padova, Italy
* Correspondence: sarah.tonello@unipd.it

Received: 9 October 2020; Accepted: 2 November 2020; Published: 4 November 2020

Abstract: Printed electrochemical biosensors have recently gained increasing relevance in fields ranging from basic research to home-based point-of-care. Thus, they represent a unique opportunity to enable low-cost, fast, non-invasive and/or continuous monitoring of cells and biomolecules, exploiting their electrical properties. Printing technologies represent powerful tools to combine simpler and more customizable fabrication of biosensors with high resolution, miniaturization and integration with more complex microfluidic and electronics systems. The metrological aspects of those biosensors, such as sensitivity, repeatability and stability, represent very challenging aspects that are required for the assessment of the sensor itself. This review provides an overview of the opportunities of printed electrochemical biosensors in terms of transducing principles, metrological characteristics and the enlargement of the application field. A critical discussion on metrological challenges is then provided, deepening our understanding of the most promising trends in order to overcome them: printed nanostructures to improve the limit of detection, sensitivity and repeatability; printing strategies to improve organic biosensor integration in biological environments; emerging printing methods for non-conventional substrates; microfluidic dispensing to improve repeatability. Finally, an up-to-date analysis of the most recent examples of printed electrochemical biosensors for the main classes of target analytes (live cells, nucleic acids, proteins, metabolites and electrolytes) is reported.

Keywords: printed biosensors; printing technologies; electrochemistry; point-of-care

1. Introduction

In recent decades, printed electronics, which include all the additive manufacturing techniques to fabricate sensors, circuits, and active and passive electronic components, has gained increasing attention due to advantages in terms of process flexibility, cost and time effectiveness [1,2]. Focusing on the biomedical area, the potential of printed electronics has recently been exploited for the fabrication of bio-sensing electrodes and their conditioning circuits. In this framework, printed electrochemical biosensors have acquired widely recognized relevance in various fields ranging from basic laboratory research to commercially available point-of-care. Thus, the possibility to obtain a sensitive analysis with a time and cost-effective approach, relying on disposable materials and on user-friendly protocols for transduction, is highly demanded by medical personnel, biologists and biotechnologists [3].

Moreover, in basic laboratory research, the possibility given by electrochemical biosensors to correlate electrical quantifiable signals with cell functions or with biomolecule/pathogen concentrations represents an interesting tool for improving the investigation of cellular pathophysiological processes and of their interaction with pathogens [4]. In hospital-based medicine, non-invasive and sensitive bio-sensing gives the possibility to improve the care of patients through ad hoc monitoring during hospitalization, contributing to better detection of bacterial infections [5], and to adjust treatment due to sensitive feedback about patient status [6]. In diagnostics, the possibility to enable the reliable detection of very low concentrations of pathology-related biomarkers, with reduced time and costs with respect to actual biochemical and molecular assays, could bring a revolution in the early diagnosis of pathologies like cancer, cardiac or neurodegenerative diseases [7,8]. Finally, the possibility to integrate those biosensors in standalone platforms (e.g., wearable, point-of-care), usable even by non-experts at home, could provide a powerful contribution to eHealth and telemedicine [9–11].

Recent advances in the development of micro- and nanoscale bio-transducers capable of detecting changes down to the molecular level, enabled by technological advances, have strongly accelerated the improvement of the metrological issues still affecting electrochemical biosensors. Those metrological characteristics encompass sensitivity (slope of the calibration plot, given by the ratio between output and input signals), selectivity (ability to correlate changes to a specific analyte, reducing the cross-sensitivity), signal-to-noise ratio (SNR, ratio between the signal of interest and background noise), repeatability (stability of the results among multiple analysis performed under the same conditions) and stability (repeatability in long-term monitoring) [12]. Another relevant useful quantity commonly adopted to compare results in chemistry/biology sensing is the limit of detection (LOD), which express the lowest quantity of an analyte that can be distinguished from the absence of that substance (a blank value) with a stated confidence level (generally 99%). It is estimated from the mean of the blank, the standard deviation (SD) of the blank, the slope (analytical sensitivity) of the calibration plot and a defined confidence factor (usually 3SD) [13,14]. It can also be considered as an indicator of the resolution of the system obtained with a statistical approach, since it is taking into consideration both the contribution of uncertainty and of resolution [13].

Looking at electrochemical biosensors from a metrological perspective, it is undeniable that their characteristics need to be discussed and compared with really competitive counterparts: mass-based and optical biosensors [15,16]. Mass-based devices also referred to as gravimetric biosensors, apply the basic principle of a response to a change in mass, using piezoelectric crystals, in the form of resonating or as surface acoustic wave devices [17]. Their main advantage is their high sensitivity to minimal mass changes, especially for molecules that are neither electroactive nor fluorescent [18,19]. Optical biosensors, both label free and label based, are based on the interaction of optical fields with biorecognition elements, showing well-known levels of sensitivity and specificity [20,21]. Despite those clear advantages and emerging trends in the area of fiber optics [22], both mass-based and optical biosensors show significant challenges in terms of their lack of repeatability, high dependency upon contour variables, high cost, high fragility, limited flexibility, and the portability and integrability of the overall readout system with more complex systems (e.g., point-of-care) [23]. Thus, compared to mass-based [24] and optical [25] biosensors, electrochemical sensors are easier to fabricate and miniaturize, facilitating the possibility of their integration on the same sensing substrate and also customized readout circuits [26]. Regarding metrological performances, despite recent advances in nanostructures, nano-printing strategies and hybrid nano-molecules that have strongly improved the LOD, the main challenges for electrochemical biosensors concern selectivity, repeatability and stability [27]. Recent advances in the area of printing technologies combined with advances in bio- and electrochemistry, nanostructures, solid-state and surface material physics, integrated circuits, microfluidics and data processing offered the possibility to address a whole new generation of electrochemical biosensors [28]; however, these biosensors require attention in relation to their metrological performance.

Compared to the most commonly adopted techniques to fabricate electrochemical biosensors, such as subtractive manufacturing, thin film, vacuum, lithography and electro-based deposition, printing technologies offer unique opportunities in terms of miniaturization, integration in complex systems and ease of customization (Table 1) [29].

Table 1. Main fabrication techniques for electrochemical biosensors: advantages and challenges (referenced articles are limited to the recent literature focusing on critical evaluation of positive and challenging aspects of the reported techniques).

Fabrication Techniques	Advantages	Challenges	Refs
Bulk Electrodes	higher stability, larger surface	no possibility of miniaturization, large volumes of sample needed, low customization possibility	[30,31]
Printing Technologies	miniaturization, low cost, wide range of inks and substrates available, integrability, complex geometries, possible combination with nanostructures, with bio-receptors	stability, repeatability, compatibility among materials	[4,8,25,32]
Thin Film (Vacuum-Based, Spin Coating)	fine control of the thickness, low costs, high repeatability	high temperatures, vacuum needed, non-compatible with low-melting point substrates, no complex geometries	[33–35]
Lithography	high resolution, high accuracy, high repeatability	long process, needed particular materials, mask based, high costs, limited available substrates	[15,36,37]
Electrospray, Electrospinning	good control of fibers, control of porosity, possibility to combine multiple materials	low lateral resolution, no complex geometries	[38–40]

The available equipment for printing technologies ranges from economic devices ensuring very low-cost production, which are ideal for rapid prototyping, to the most expensive ones providing a greater geometrical resolution, which are in some way comparable with standard lithographic methods, but without the need for clean rooms and/or multiple step processes with sacrificial layers [2,25,41]. Overall, the printing technologies employed for fabricating electrochemical sensors can be classified between contact printing (gravure, flexographic, offset, micro-contact dispensing and screen printing (SP)) and non-contact printing (inkjet (IP), aerosol jet printing (AJP), laser-induced forward transfer (LIFT), micro and nano-pen printing). Contact printing encompasses all the mask-based techniques in which patterned structures with inked surfaces and substrate are in physical contact. These techniques ensure high throughput and thus are often (e.g., SP) the most frequently adopted for low-cost and rapidly fabricated biosensors [42]. However, since they are characterized by high material waste, limited resolution and a limited range of materials (substrates, inks and solvents), increasing attention has recently been paid to non-contact printing techniques (also defined as maskless techniques). These technologies are based on ink dispensed through openings or nozzles and define structures by moving the stage in a pre-programmed pattern. Thus, they allow for a reduction in material waste, the simplification of the printing process, an improvement in its control and flexibility and also enable improved resolution, miniaturization and more complex patterns (Figure 1) [43,44].

Along with the advantages discussed, challenges in terms of compatibility among the wide variety of materials used in the fabrication of sensors represent a predominant issue that must be faced to ensure the feasibility and metrological performances of the printed devices. The most recent emerging non-contact techniques [46] are aiming to optimize the processes of ink deposition, reducing the dimensions of droplets (micro- or nano-pen printings [47,48]), through the finest control of printed track width using lasers (LIFT) or by focusing aerosol ink through a stream of gas (AJP) [49]. Additionally, novel sintering methods (e.g., photonic curing) are under investigation to optimize ink post-processing. These emerging techniques are thus trying to face the challenges in terms of conductivity, repeatability and standardization that are still openly affecting printed biosensors when compared with their bulk counterparts [50]. Additionally, the possibility to combine and customize different materials and to exploit novel curing methods with respect to other traditional techniques (e.g., laser cutting, machining) opens the way for the effective integration of biosensing with directly printed microfluidic circuits (e.g., paper based, polymer based) and embedded electronics (insulating layer and conductive tracks), with consequently improved costs and time effectiveness [4,9,51].

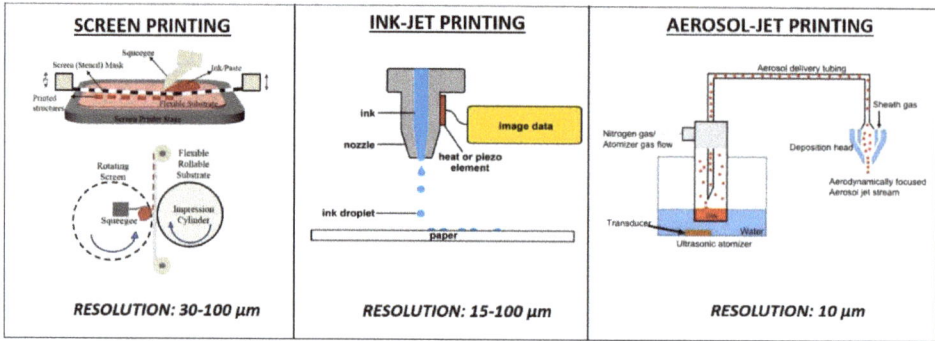

Figure 1. Comparison among fabrication processes to print electrochemical biosensors, in terms of ink dispensing and resolution achieved. Reproduced with permission according to the terms of the Creative Commons Attribution 3.0 license from [43–45].

Considering this, the aim of this review is to provide an up-to-date picture of the state of the art of printed electrochemical biosensors. First of all, this paper focuses on the opportunities offered by printing technologies for electrochemical biosensors in terms of transducing principles. Following this, a discussion on the main metrological challenges of printed electrochemical biosensors is performed. In particular, we focus on how enhancing the printing approach, combined with the most innovative technologies in terms of nanostructures, microfluidic and non-conventional substrates, is opening up promising avenues through which to face those challenges. Finally, a review of the most upcoming trends of printed biosensors for the main target analytes (cells, nucleic acids, proteins, metabolites and electrolytes) is provided.

2. Transducing Principles of Printed Electrochemical Biosensors

The transducing principles of printed electrochemical biosensors can be grouped into three main classes: amperometric, impedimetric and potentiometric [52]. Common advantages brought by printing technologies to all three classes are related to the miniaturization of the electrodes, to the use of nanostructured inks, to printed microfluidic paths and via the extension to non-conventional substrates.

Thanks to printing technologies, both three-electrode (for amperometric) and two-electrode (for impedimetric and potentiometric) conformations, traditionally implemented with solid electrodes in a baker containing several milliliters of samples, can be easily miniaturized onto a small substrate, ensuring a reduction in the required sample volume from milliliters to a variable range within picoliters and microliters [49]. Moreover, the capacitive background current associated with the charging of the double layer is reduced proportionally to the reduction in the surface area of the conductive electrodes. The resistive drop in the electrode–solution system is reduced by shortening the ionic current path in miniaturized cells. Overall, those elements contribute to reducing the interfering noise coupled to the electrodes. The reduced time constant coming from reduced capacitance and resistance enables faster electron transfer kinetics monitoring.

Printing technologies enable an easier fabrication of microfluidic circuits. This possibility, combined with high-resolution nanostructured coatings, enhances the accuracy and sensitivity. In fact, thanks to the high accuracy of sample delivery to the sensing area and to the presence of nanowires and nanospheres, the interaction between the analyte and the electrode active area is enhanced, changing it from a 1D planar diffusion to a more uniform 2D or 3D diffusion. The use of nanoinks allows to increase the surface to volume ratio, increasing the active area useful for redox current detection, for impedance variation or charge accumulation detection, bringing an improvement in terms of overall sensitivity. Furthermore, the highest control obtained in these microsystems in terms of sample dispensing, ink and coating deposition can also improve the repeatability of the electrochemical measurements [53,54]. Overall, the combination of the reduction in the interference noise processes and the enhancement of the transducing effect of the measurand achievable in printed miniaturized integrated biosensors increases the signal-to-noise ratio of such bioanalytical systems [55,56].

In addition to working electrodes (WE), the potential of printing techniques also needs to be exploited for improving counter (CE) and reference electrodes (RE), which require particular attention when aiming for electrochemical cell miniaturization [57]. CE represents the element required to complete the circuit with the WE, thus allowing the charge coming from the reaction on WE to flow and be read [58]. Consequently, its size should be much larger than the WE to ensure no current limitations arise. Thus, nanostructures and complex geometries made available by emerging printing are under investigation to increase the surface to volume ratio and to guarantee proper control of the electrical parameters of the cell during the analysis [59]. Regarding RE, it is the element that needs to be kept at a constant potential during all the analyses, to control the potential of WE (e.g., in voltammetry) or to allow measurement of an indicator electrode (e.g., in potentiometry). Thus, attention is being paid to novel materials and curing strategies to improve the stability of RE and limit the influence of surrounding conditions [60].

Despite these common advantages, due to significant differences in terms of speed, sensitivity and selectivity among amperometric, impedimetric and potentiometric biosensors, the specific potential offered by printing technologies for each class needs to be discussed, considering their intrinsic characteristics (Table 2) [15].

Table 2. Review of main advantages and challenges of the three main groups of electrochemical techniques (referenced articles are limited to the recent literature focusing on critical evaluation of positive and challenging aspects of the reported techniques).

	Detectable Analyte Concentration	Advantages	Challenges	Ref
Amperometry/ Voltammetry	lower than 10^{-12} M	highest sensitivity, high specificity, continuous monitoring, possibility to detect many compounds with different characteristic potentials in one measurement	required electroactivity, current production, interferences, effect of surrounding environment, long-term stability (degradation of materials or of labels), time-consuming	[49,61–63]
Impedance spectroscopy/ Conductometry	~10^{-8} M (some recent example down to ~10^{-12} M)	miniaturization, limited invasiveness, several information frequency-dependent, direct real-time monitoring, no references electrode needed, no need for redox probe (label free)	need nanotechnologies to improve sensitivities, potential error due to double layer capacitance of non-target analytes, intrinsic non-specificity, mathematical modeling needed to extract information	[53,54,64–66]
Potentiometry	~10^{-8} M	simple conditioning, miniaturization, real-time monitoring, no current flowing, limited invasiveness, no electroactivity required	intrinsic non-specificity, very sensitive to temperature changes, possible ionic buffer interferences, frequent recalibration needed	[62,67–69]

Next, the basic working principles, advantages and disadvantages of amperometric, impedimetric and potentiometric biosensors will be overviewed, focusing on the specific contribution and improvement brought by the printing approach to each method. For more extensive and theoretical details of each electrochemical technique, out of the scope of this review, we suggest that the reader deepens their knowledge of this theoretical topic in the related literature [61].

2.1. Amperometric

In amperometry, a three-electrode conformation is used, comprising a WE, CE and RE electrode. WE potential is controlled through a signal from a generator and the current resulting from the oxidation and/or the reduction reaction of electroactive molecules exchanging electrons with the WE conductive surface are then measured in the loop closed by the cell. If the signal coming from the generator is varied, then the methods belong to the sub-class of voltammetry [61].

The main challenges of printed amperometric biosensors still refer to cross-sensitivity, the interferences of the buffer composition and the effect of the surrounding environment [70]. Concerning the influence of contour variables, the most challenging aspects refer to interfering molecules (inks, mediators, labels) with similar potential. Concerning implantable electrodes or analyses performed on biological fluids, a relevant issue is the electrode fouling by non-target proteins and biomolecules, which can limit direct electrode exchange. Furthermore, the accuracy and stability of the currents measured are particularly challenging for both short and long-term measurements [71]. Static measurements, in the absence of stirring and without proper fluidics, can be easily affected by saturation due to species accumulation, by difficult low current detection due to double-layer capacitance or by a decrease in electrode performances due to the degradation of ink or of the ink–substrate bonding [72].

A smart combination of high-resolution nanostructure direct printing with peculiar techniques able to enhance low faradaic currents and not background processes (e.g., differential pulse voltammetry) can help to face those issues, reaching LOD < 10–12 M, the lowest among electrochemical techniques [52]. Finally, focusing on biosensor selectivity, cross-sensitivity of different species can be improved thanks to the flexibility in ink preparation. The possibility to directly print selective electroactive labels allows to enhance the selectivity of currents resulting from voltammetries using nanoparticles and nanostructures as electroactive labels (limiting the need for additional markers) [73] and to improve repeatability due to better control of the deposition process.

2.2. Impedimetric

Impedimetric biosensors are based on the direct correlation of impedance changes with changes in terms of target analyte concentration, without requiring additional labels or biomolecule electroactivity. After applying an alternate voltage to the two electrodes (WE and CE), with a constant amplitude (usually between 5 and 10 mV) and a defined frequency range-, the resulting alternate current is measured and the overall impedance (Z) correlated with analyte concentration [69]. They provide the result directly, without requiring the electroactivity of the target analyte. Impedimetric biosensors based on the principle that biomolecules bound onto a printed conductive surface are acting as insulators (e.g., adherent cells proliferation monitoring) fall in the subclass of reactive [74]; the ones based on the measurement of electrolytic conductivity to monitor the progress of a chemical ionic reaction instead fall in the class of conductometric [68].

Among the most important advantages of impedimetric biosensors compared to other classes are the low voltages employed, which do not damage or disturb most bio-recognition layers [75]. From the point of view of the target analyte, the small excitation signals adopted cause small amplitude perturbations from the steady state, which makes this method optimal to monitor in real time the dynamics of biomolecule interactions and the pathophysiological processes of living cells, without significant alterations to the ionic balance in the extracellular space [53,54,76]. Furthermore, from the point of view of materials, this low invasiveness gives the possibility to explore novel non-conventional organic conductive materials (e.g., conductive functionalized polymers or small molecule organic semiconductors) with peculiar surface modifications that can enhance the sensitivity and the LOD of the analysis [77].

The challenging aspects of impedimetric biosensors are the strong influence of pH, temperature, buffer characteristics or non-reacting ions on measurement accuracy and repeatability [76,78], the worse detection limits compared to potentiometric or amperometric methods (usually around 10^{-8} M) and the sources of error due to double-layer capacitance and electrode polarization [68]. Furthermore, the wide spectrum of frequencies of the applied voltage implies a very small power at each frequency and, consequently, a limited SNR of the impedance measurement with respect to other electrochemical techniques [79].

The opportunities of the printing approach for impedimetric biosensing mainly refer to the possibility to exploit novel nanostructured inks to enhance SNR and to the availability of biocompatible organic inks to improve the integration of sensing elements in biological environments. Thus, due to the limited invasiveness of the technique, printable organic and degradable inks can also be deposited on the electrode to investigate live cells, allowing impedimetric monitoring during a long-term culture both in 2D and 3D environments [80,81].

2.3. Potentiometric

In potentiometric biosensors, the measurement is performed in zero-current conditions, with a two-electrode structure, without the need for a generator or current measurement device. The voltage across WE and RE is measured with a high-input impedance device, to minimize the contribution of the ohmic potential drop to the total difference in potential. The potential of WE, thanks to an accumulation of charged molecules (ions), exclusively depends on the analytical concentration of the analyte in the gas or solution phase, while the RE is needed to provide a defined reference potential [62].

Those biosensors can be easily miniaturized and integrated in all printed devices since they require low-cost measurement instrumentation. Due to the simple electronic conditioning circuit, potentiometric biosensors show a rapid response, ease of use and robustness. On the contrary, their main intrinsic challenges are related to their non-specificity, to the influence on temperature variation, to the need for frequent re-calibration and to false positives due to interfering charged molecules in solution [60,61,82].

Thanks to the progress of additive manufacturing, printed potentiometric biosensors are undergoing a renaissance, with improvements in the detection limits (down to $\sim 10^{-8}$ M) and selectivity enabled by the introduction of novel materials and the integrability of these sensing concepts with wearable and implantable devices [67]. The possibility to fabricate miniaturized electrodes with customized inks could provide improvements in terms of the stability of RE, tuning the ink composition [57,60] and the selectivity of the approach, directly printing selective coatings to substitute for the selective membranes that are traditionally adopted [83]. Other great opportunities provided by potentiometric measurements combined with printing technologies refer to the possibility to realize innovative sensors on degradable or biological substrates (directly on the skin or implanted in the human body) due to the sensing principle at zero current, which limits the possible perturbation in the sensing area [84].

3. Discussion of Opportunities of Printing Technologies and Metrological Challenges of Electrochemical Biosensing

The metrological characteristics of electrochemical biosensors represent the main challenges still slowing down their maturity for robust comparisons within different scientific experimental results and for final reliable use in clinical settings, laboratories and point-of-care applications. Thus, the high sensitivity, low uncertainty, high repeatability, low cross-sensitivity of environmental influence and long-term stability are all essential requirements to performing a meaningful comparison between different repetitions of the same experiments not only at different times, but also within different laboratories [49,85]. Considering the framework described, printed electrochemical biosensors still present large areas for improvement in terms of both metrological performances and application conditions. In these terms, the main opportunities of the printing approach are discussed as powerful tools to improve the metrological performances of biosensors not only in terms of LOD, sensitivity, selectivity, repeatability and stability, but also to enlarge their field of application in environments with non-optimal working conditions (e.g., high humidity, salinity or biologically degradable environments).

3.1. Printed Nanostructures to Improve LOD, Sensitivity and Repeatability

Printable inks offer interesting opportunities for customization due to the wide variety of nanostructures and biomolecules that can be incorporated. As highlighted by recent research [86,87], even with the same chemical composition, electrode superficial nano-structuration strongly influences the properties of the finally fabricated biosensors, in terms of both LOD and sensitivity, due to the increase in the active area available for interaction with nano-molecules [88]. Additionally, the possibility of achieving a uniform distribution of nanostructures through the use of multiple supporting printable materials, and of improving the orientation of nano-molecules (e.g., DNA, RNA, antibodies, aptamers) thanks to nano-printing methods [89,90], can have relevant impacts on improving the effectiveness of nanostructure–biomolecule interactions, with the consequent enhancement of the sensitivity, repeatability and LOD of the measurement (Examples in Figure 2).

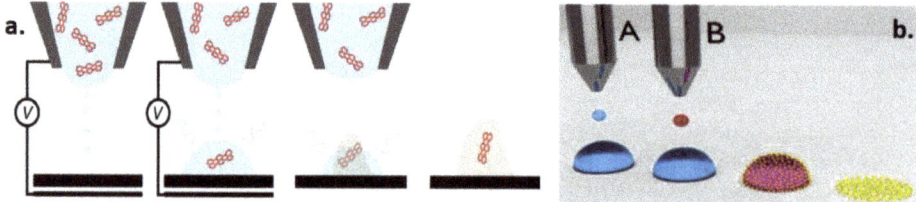

Figure 2. Two examples of strategies to enable finest control of nanostructure printing: (**a**) a schematic of the main steps of on-demand electrohydrodynamic dropwise deposition, solvent evaporation and crystallization, capturing a single molecule in the crystallized deposit and thus achieving oriented nano-molecules [89]; (**b**) how two-step printing strategies with supporting printable materials can help to enhance the uniformity of printed nanostructures [90]. Figures reproduced with copyright permission from John Wiley and Sons [89,90].

The use of printed nanostructures was demonstrated to improve the quantifiable LOD, from µM levels commonly observed with bulk electrodes down to nM or even pM levels (lower than traditional gold standard techniques such as ELISA) [46,51]. LOD improvement was shown to be strongly dependent on the type, size and composition of the nanostructures, and to be enhanced when relying on a combination of different nanostructured materials [91]. Furthermore, most of the leading research in electrode nano-structuring has recently confirmed that an accurate micrometric control of nanostructure deposition onto electrodes through micro and nano-printing strategies also represents a winning strategy to lower the relative standard deviation of the overall measurement (<5.0 % compared with the common relative standard deviation (RSD) of 20% registered in electrochemical sensing without control of surface material deposition) [46]. Increasing attention has recently been addressed to the investigation of novel materials and shapes that improves the metrological aspects of LOD, SNR and sensitivity. The use of nano-cubes of novel graphene-based nanostructures, realized with a combination of different materials [92], to enhance cell–biosensor interaction [93], and of printed nanostructures combined with novel curing techniques, have been highlighted in the recent literature as promising to improve the performance of paper-based biosensors [94]. Furthermore, in [95,96], specific comparisons in terms of sensitivity were performed among carbon nanotubes, as well as fullerene and platinum printed nanostructured electrochemical sensors, demonstrating the combined effect of the chemistry, shape, dimension and deposition techniques of the nanostructures on LOD and repeatability. An improvement in the LOD in quantifying IL-8 (from 2 ng/mL to 0.38 ng/mL) and p53 proteins (from 2 ug/mL to 100 ng/mL) could be obtained with nanostructured biosensors with respect to their non-nanostructured counterparts. Interestingly, in [97,98], carbon nanotubes and other functional nanomaterials were shown as also being useful to improve the SNR of electrochemical techniques, since they can, on the one hand, provide excellent electrical conductivity and promote radial diffusion and, on the other, reduce the area of double-layer capacitances. Finally, in [99], comparing different nanostructure deposition strategies while quantifying the very same protein with the same protocol, direct nanostructure printing through AJP deposition was demonstrated as the most sensitive and reproducible technique. It was thus demonstrated that a higher spatial accuracy in the deposition of nanostructures brings improvements both in terms of LOD (improved from (LOD from 2.1 to 0.3 ng/mL) and in the relative standard deviation (RSD, reduced from 50% to 10%), with promising results possibly extended to electrochemical sensors for several diagnostic and medical applications.

3.2. Printing Strategies to Improve Organic Biosensors Integration in Biological Environments

One of the areas that has recently gained much attention is the use of biosensors directly embedded in biological environments (implanted in the human body or integrated in cell culture) to obtain reliable feedback from biosensors. In addition to printability and biocompatibility, essential requirements in these applications relate to the adaptability of biosensor elements (e.g., inks, coatings or conditioning circuits) to an environment traditionally harsh for electrical instrumentation, with high humidity and salinity at physiological temperature (around 37 °C).

In this framework, despite the fact that inorganic materials would be commonly preferred due to their higher stability and metrological performances, in the recent literature, growing interest has been addressed to the use of organic printable materials due to their higher biocompatibility and non-invasiveness. In particular, conductive polymers [100], carbonaceous materials [101] and organic semiconductors (poly(3,4-ethylenedioxythiophene) polystyrene sulfonate (PEDOT:PSS), Triisopropylsilyl ether (TIPS)) [102–104] all represent attractive candidates due to their low cost, good compatibility with most of the printing process and customizable chemical composition [105].

Despite the fact that several examples have been proposed in areas ranging from cell monitoring to implantable devices [105,106], the main metrological challenges refer to repeatability (often higher than 10%), SNR (often lower than 20), due to intrinsic variations in the background impedance, and the stability of the electronic performance over long periods (most of the works demonstrated only a few days, while feedback on cell cultures would be interesting over longer periods of a few weeks) [97,107]. Only facing those metrological challenges can ensure the intra- and inter-laboratory repeatability required for biosensor validation, opening the way to a whole new world of biosensing that is more biomimetic and integrated with living environments [108].

To this end, specific attention has recently been addressed to exploiting the potential of printing technologies for customizing ink preparation, deposition and curing for both electrodes and coatings [33].

Regarding ink customization, the possibility to tune the ink chemical composition of conductive polymers (e.g., PEDOT:PSS, polyaniline, TIPS-pentacene) represents a powerful strategy that could lead to controlling the metrological performance of biosensors through a finer control of electrode material solubility and degradability, in agreement with target analyte dynamics [109]. Promising examples have demonstrated conductive inks for embedding sensing elements into the human body or 3D scaffolds [110,111], or investigating organic semiconductors (TIPS-pentacene) to realize transistors for monitoring neural cell culture activities, due to their combined printability, biocompatibility and degradability [112].

Another opportunity of the printed approach refers to the possibility to directly print customized coatings onto conductive electrodes. Moreover, an improvement in terms of stability can be brought by printing protecting material (e.g., UV-curable polymers or dielectric layers) to avoid direct contact with ions or water. Furthermore, promising results in terms of repeatability were obtained with the micro and nano-printing of enzymes [113], proteins [114] or cells into scaffolds [115].

Regarding stability and SNR during long-term cell monitoring, increasing attention has also been recently addressed to nanostructures and to emerging ink deposition and curing (e.g., AJP) to improve the stability of an effective ink–substrate interaction and consequently of the metrological performance [80,116]. A promising strategy to enhance SNR demonstrated the introduction of carbonaceous nanostructures and the use of emerging micro- and nano-printing techniques to enable their uniform distribution [117,118]. Interesting results from [97,119] showed a five-fold improvement in the standard SNR. Among the emerging technologies, AJP was shown to be effective in fabricating printed carbon electrode that were integrable with glassware as modular systems to monitor the growth and differentiation of human colorectal adenocarcinoma cells (CACO-2) in static 2D cultures [116], and the proliferation of mesenchymal stem cells into 3D scaffolds [80], with stable performances over 21 days of culture (Figure 3). Furthermore, contactless technologies combined with proper surface treatment were also demonstrated to achieve stability in organic carbon inks during dynamic myocyte 2D cultures using stretchable substrates, with a sensitivity of 80 Ω/cell) and a RSD around 20% [120].

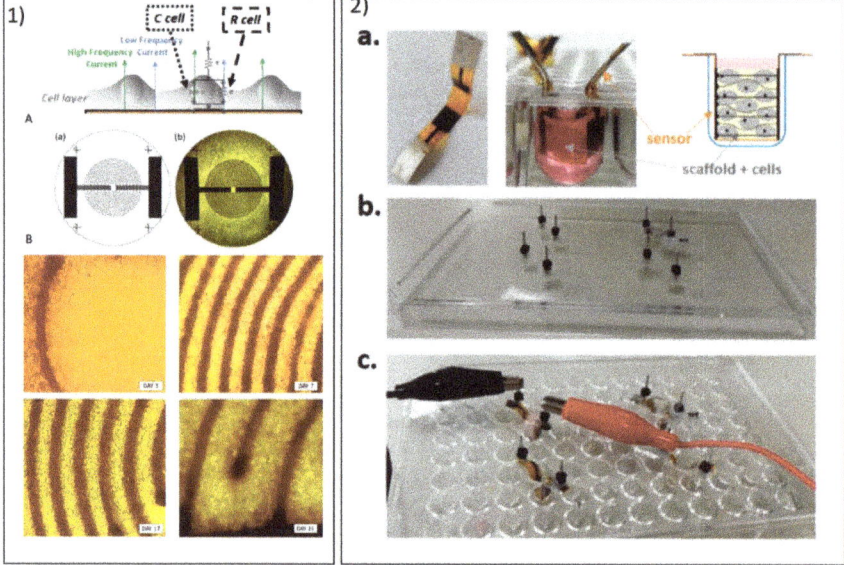

Figure 3. Example of how aerosol jet printing (AJP) biosensors fabricated with organic carbon-based ink designed to enable long-term noninvasive monitoring of cell cultures: (**1**) example of interdigitated carbon-based electrodes customized for multi-well plates for 2D monitoring of the differentiation of CACO-2 cells [116]. Reproduced with copyright permission from Elsevier. (**2**) The set up proposed to monitor mesenchymal stromal cells through foldable parallel carbon electrodes directly within 3D scaffolds [80]. Reproduced from an open access publication.

3.3. Emerging Printing Technologies for Non-Conventional Substrates

Printing technologies allow for the exploitation of a wider variety of substrates compared to traditional techniques. In recent years, several emerging methods have been proposed, enabling greater control of multiple degrees of freedom and also droplet dimension, with a direct effect on resolution with respect to traditional techniques such as SP or IP. These techniques (e.g., micro and nano dispensing, AJP) allowed researchers to enlarge the substrates available from traditional rigid and planar substrates (e.g., ceramic or silica) to non-conventional substrates (e.g., plastic, paper or stretchable substrates). Among non-conventional substrates, paper-based substrates, stretchable and 3D substrates represent the most investigated ones to enable the integration of sensing elements into disposable devices, into substrates undergoing mechanical stretching and on the irregular surfaces of complex structures [121,122] (Figure 4).

Despite clear advantages over rigid traditional substrates, the efficiency of the production and the ease of processing on those non-conventional substrates must be improved before allowing commercialization [123]. Several metrological challenges relate to the performance in terms of the repeatability (with an RSD higher than 10% preventing commercialization [124]) and sensitivity of electrochemical biosensors realized on non-conventional and 3D substrates. In this way, particular attention to the electronic transducing aspects of fully printed devices onto non-conventional substrates have been placed under investigation to correlate the response of fully printed devices with the specific properties of the printed material and on the geometrical characteristics of the electrodes [11].

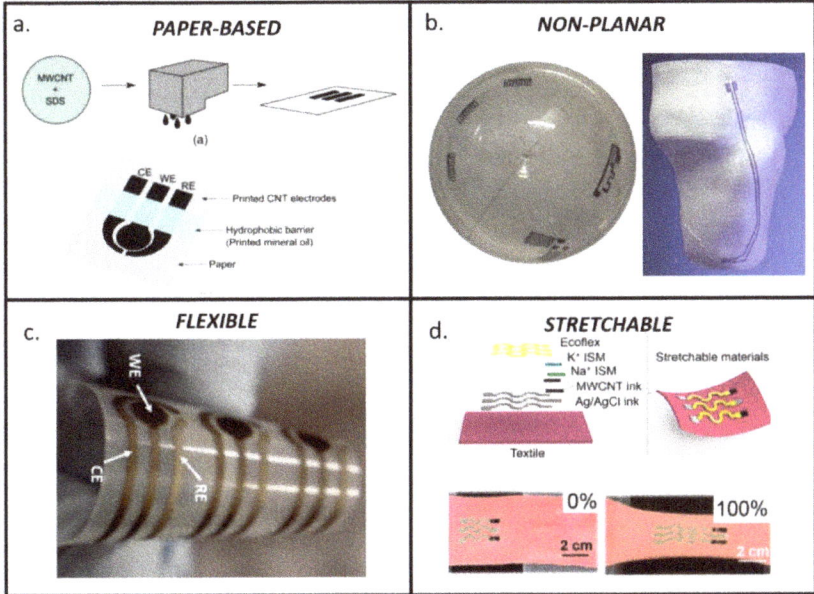

Figure 4. Summary of the main classes of non-conventional substrates enabled by printing technologies: (**a**) paper-based biosensors, often enhanced by nanostructured, as reviewed in [125]; (**b**) biosensors printed on non-planar surfaces, examples presented in [126,127]; (**c**) example of three-electrodes layout for histamine detection printed onto a flexible substrate, [128]; (**d**) a recent example of electrolyte detection for printing electrochemical sensors for wearable applications onto highly stretchable substrates, reproduced by [129]. All figures were adapted from open access papers cited under the Creative Commons license.

Novel emerging methods for ink dispensing and curing are trying to face these challenges by improving the performances of biosensors realized on unconventional substrates. In particular, novel, non-contact printing techniques (e.g., AJP, nano dispensing) can achieve resolutions of a few micrometers, along with very good accuracy and repeatability even on materials with poor porosity or with irregular surfaces [46,130]. Regarding the use of stretchable substrates, increasing attention has also been recently addressed to the opportunities and limitations of stretchable inks and substrates [131], highlighting that both geometry optimization and perfect matching between stretchable substrates and inks should be carefully addressed in order to guarantee an optimal performance during all the different phases [132].

Regarding the disposability and optimization of cost effectiveness, paper represents the most promising substrate to combine cost effectiveness with intrinsic capillary properties in order to improve sample flow control [125,133–135]. Furthermore, a paper substrate can truly provide environment friendliness for electrochemical sensors, making them disposable while respecting the standard of the green era and the circular economy [136]. Paper has been used for more than a century in analytical and bioanalytical devices and, nowadays, recent advances in developing paper-based immunosensors, aptasensors and genosensors are highlighted as very promising solutions that combine sensitivity with low cost and disposability [137]. The high performance reached in terms of ink deposition by techniques such as micro-dispensing or AJP is paving the way to the fabrication of low-cost disposable biosensors with metrological accuracy, repeatability and stability comparable with their traditional counterparts [137,138]. Interesting examples are under investigation in order to achieve a combination of biosensing elements [139] and complete circuit fabrication [140] onto cellulose substrates, attracting attention for enabling smart food monitoring into disposable paper-based packaging. Resistivity values of 26.3×10^{-8} $\Omega \cdot$m on chromatographic paper, 22.3×10^{-8} $\Omega \cdot$m on photopaper and of 13.1×10^{-8} $\Omega \cdot$m on cardboard were obtained by AJP. These values are comparable with the range of resistivities obtained with similar inks on conventional substrates (from 4×10^{-8} $\Omega \cdot$m to 44×10^{-8} $\Omega \cdot$m depending on deposition and curing parameters) [140,141]. This represent a promising result for integrating electronic tracks on disposable substrates for food packaging, wearables or point-of-care. Furthermore, a combination of paper-based substrates with nanostructures, with origami architectures and with sensitive electrochemical techniques such as Differential Pulse Voltammetry (DPV) or Anodic Stripping Voltammetry (ASV), is enabling researchers to reach a competitive limit of detection in the order of fM [142,143] (Figure 5).

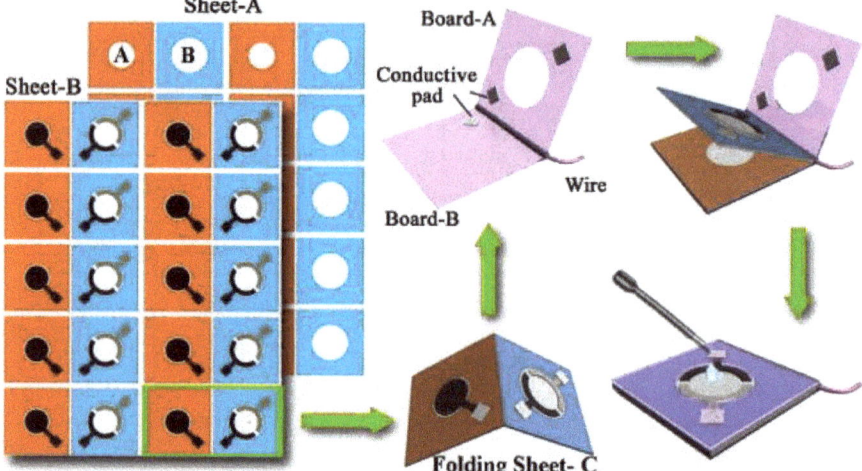

Figure 5. Interesting example of paper-based printed origami biosensors: after multi-plane printing, electrode folding ensures better control of the sample and higher repeatability of the measurement. Reproduced from [143] with copyright permission from Elsevier.

Regarding irregular surfaces, micro and nano-dispensing printing techniques are also a great opportunity for producing biosensors directly onto 3D surfaces [144,145]. These emerging methods, in addition to an optimal control on multiple degrees of freedom, allow for rapid and more effective ink drying, sintering and curing over a wide range of substrates, aspects that are required to improve ink adhesion on irregular surfaces, as well as its conductivity and stability [138]. Thus, differently from flat 2D surfaces, when printing on surfaces with high inclination and rugosity in addition to standard post-printing curing, particular care should also be addressed to primers or to ink drying and/or polymerization during printing, to enable the optimal adhesion of ink on those surfaces [126,127,146]. This provides the possibility to directly integrate repeatable and stable sensors onto already-fabricated products (as in Figure 4b), without the need for attaching external electrodes [99,147,148], but also to integrate highly conductive printed tracks of customized conditioning electronics in an all-in-one structure using a single fabrication technique [149]. This represents an advantage, both for point-of-care applications and for wearable devices, to limit the obtrusiveness during a long-term recording of the patient.

3.4. Microfluidic Dispensing to Improve Repeatability

Printing technologies offer unique opportunities in terms of biosensor integration with customized microfluidics, with embedded conditioning electronics or with multisensory platforms. This allows to take a step forward from printed biosensors to standalone printed biosensor platforms, thanks to optimal process flexibility and to the wide range of materials available [150].

An aspect of predominant relevance for achieving accurate biosensing on miniaturized electrodes and for continuous analysis is to ensure proper sample management and control. Three-dimensional printing strategies could create a revolution in this sense, providing the possibility to directly realize both microfluidic systems for sample preparation/distribution and conductive electrodes within the same printing session. This represents a great improvement in terms of lowering circuit fabrication costs and the time for complete platform production. Two main methods to manage and control the sample under analysis are under investigation: (i) the fabrication of support-free microfluidic circuits on the same sensor chip; (ii) the exploitation of the peculiar capillarity of the substrate, such as paper. The first category refers to polymer-based channels that can be fabricated using UV-curable materials. Interesting examples have been shown not only for sample distribution [113], but also in combination with novel nanomaterials to incorporate the filtration, concentration and amplification of the analyte directly within the chip before reaching the measurement point [151]. The second category refers to the fabrication of lateral-flow paper-based assays, in which the paper capillarity is exploited to guarantee the efficient flow of the sample, which is better controlled with customized hydrophilic paths printed with wax or other hydrophobic materials on paper substrates [45,152]. The choice of one with respect to the other category mainly refer to the requirement in terms of the accuracy of fluid control, of the scalability of the device and of the cost of the fabrication.

These opportunities, ensuring an efficient and controlled delivery of small sample volumes, could represent a key element to increase repeatability among different batches and laboratories. Furthermore, realizing a proper microfluidic circuit able to distribute equal parts of the sample on multiple sensing areas could help to improve not only the repeatability, but also the sensitivity of the overall analysis. An example of how technologies for printed electronics are playing a relevant role in facing those metrological challenges can be found in [113] (Figure 6). The AJP strategy was applied therein to improve the repeatability and sensitivity of glucose sensing. Through a single printing process, a complete platform was developed, including the microfluidic circuit, the electrodes and enzyme-based electrode functionalization. This example appears to be particularly appealing, both in terms of cost and time effectiveness, reducing the number of materials and techniques required to get to the final results, and also in terms of repeatability, limiting the error introduced by manual sample delivery. The LOD = 2.4 mM, sensitivity = 2.2 ± 0.08 µA/mM and RSD lower than 8% confirmed the effectiveness of AJP to realize a fully printed platform and of the sum of a single well in contributing to the enhancement of the overall sensitivity in a clinically relevant range (3–10 mM). Other interesting examples of fully printed electrochemical biosensors integrated into a microfluidic structure can be found in the recent literature [153], where researchers try to improve the LOD and the repeatability of point-of-care devices for biomarker detection while, at the same time, enabling a low cost and disposability [154].

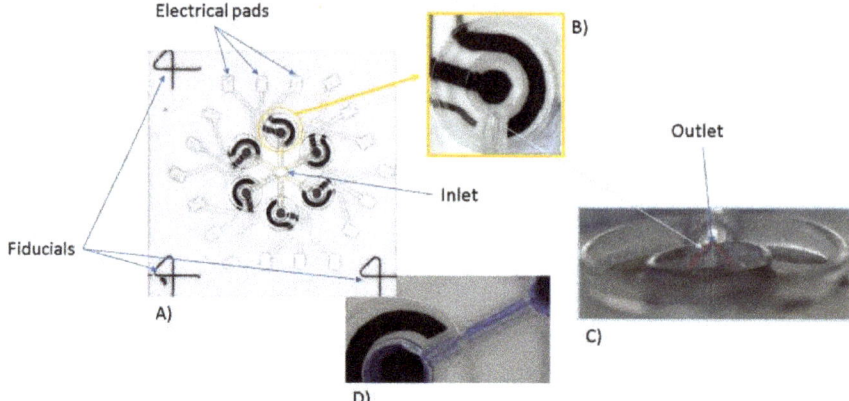

Figure 6. Example of use of fully printed integrated biosensors and microfluidic circuit realized with AJP. The figure represents the platform realized in [113], in which all the elements (electrodes, conductive tracks and polymer-based microfluidic channels) were fabricated and fully printed with the AJP technique. In details: (**A**) Layout of the complete platform; (**B**) Zoom of a single electrochemical cell; (**C**) Detail of microfluidic inlet; (**D**) Example of liquid control in each sensing point. This figure was reproduced from an open access publication [113].

4. Opportunities of Printed Approach for the Main Classes of Bio-Analytes

All the opportunities and metrological challenges of printed electrochemical biosensors discussed up to now need to be carefully considered, taking into consideration the specific target bio-analytes of interest: live cells, nucleic acids, proteins, metabolites and electrolytes [73,155] (Table 3). Next, we provide an overview of the most recent and relevant opportunities and trends that printed technologies are making available for each class of bio-analytes.

Table 3. Review of the main advantages, challenges and trends of the main target analytes for electrochemical printed biosensors (referenced articles are limited to the recent literature focusing on critical evaluation of positive, challenging aspects and trends of each class of analyte).

	Advantages	Challenges	Main Trends	Ref
Whole Cells (Eukaryotic and Pathogens)	direct detection without need for sample pre-treatment to extract and purify sample, long life-time, higher stability during time	low selectivity, challenging the detection with high sensitivity, risk of contamination, often slow reactions	organic printed biosensors, degradable sensing elements, sensors integrated in glassware and scaffolds, use of disposable non-conventional substrates, use of nanostructures to enhance sensitivity	[5,93,156–158]
Nucleic Acids	wide range of application, high specificity	needed labels, time consuming because of purification step required, high costs	nanostructures, nano-hybrid materials, combine amplification techniques with the electrochemical detection	[159–163]
Proteins	simplicity, broad spectrum of applications, well-known structure, small dimensions, sensitivity, broad range of available recognition elements with high selectivity and strong binding interaction, ease validation	poor chemical, thermal and pH stability, risk of degradation due to substrate–protein interaction, high costs of antibodies for ensure selectivity, immunogenicity	low-cost disposable materials, simplify protocols, use of direct biomolecules printing, imprinted polymers, composite materials	[164–166]
Metabolites and Electrolytes	indirectly correlated with a plethora of physio-pathological processes, detectable in multiple body fluids, ideal for non-invasive continuous monitoring of health	long-term stability of enzymes, interferences of charged non-target analytes	novel selective materials, improve integration of sensors and microfluidic circuit	[167–169]

4.1. Cells and Pathogens

Process flexibility and the low cost of the fabrication of emerging printing strategies combined with the non-invasiveness of most of the electrochemical techniques make printed electrochemical sensors ideal for cell monitoring and pathogen detection [170]. When dealing with a live target, a primary issue for any inks and substrates becomes biocompatibility. Thus, an optimal interaction between cells and substrates is fundamental to ensuring effective sensing, even before electronic performances. An additional concern is related to the high humidity and salinity that printed biosensors need to undergo when inserted into a typical cell culture environment, in samples with pathogens or those implanted in a human body. Particular attention has been recently paid to improving the reliability and standardization of the outcomes of cytocompatibility tests to support researchers during the design of printed biosensors [171–173].

Printed electrochemical biosensing of whole cells represents a useful tool to merge the advantages of electrochemical techniques for cell monitoring with the opportunities offered by printed approaches [174]. In particular, impedance spectroscopy, due to its non-invasiveness, intrinsic label-free protocol and possibility to be applied both in 2D and 3D strategies, is one of the most widespread for these target analytes [156]. The most updated examples of cell monitoring are trying to exploit 3D printing strategies to enable the integration of sensing elements with devices capable of providing mechanical and chemical stimuli to live cells [175], and to combine imaging with electrical monitoring.

To deal with the detection of prokaryotic cells (virus, bacteria), often present in concentrations lower than fM, highly sensitive voltammetries are required. Techniques such as differential pulse or anodic stripping voltammetry are able to enhance the analyte contribution against a noisy background, reaching limits of detection in the order of fM [176–178], taking particular advantage of the innovative opportunity of directly printing nanostructures or biomolecules with highly controlled coatings. Other interesting recent examples also demonstrated how printed potentiometric sensors represent a reliable tool to quantify the presence of bacterial cells forming biofilms on medical surfaces, thanks to the negative correlation between the open circuit potential and the amount of bacteria [179].

4.2. Nucleic Acids

The possibility to develop low-cost, sensitive and rapidly printed biosensors for nucleic acid quantification is of particular interest in the field of diagnostic and screening tests, since those targets are typically key indicators of cells, viruses and bacteria, and are often responsible for pathological conditions [180]. Thus, the development of printed devices that are usable outside hospitals and laboratories is highly requested to limit pathological spread and/or to optimize clinical management, as recently strongly highlighted by the pandemic due to SARS-COV-2 [181]. However, if, on the one hand, the design of reliable and competitive printed electrochemical biosensors for those targets is particularly attractive, on the other hand, it is also very challenging due to the high standards offered by the currently adopted molecular techniques (e.g., polymerase chain reaction (PCR)). Thus, traditional techniques are affected by long processing times and the high cost of instrumentation and reagents, but they remain the current gold standard, since they are the only techniques able to reach LODs down to few copies of DNA/RNA [182]. The enormous advances in terms of nanostructure–DNA hybrid structures, ultra-high-resolution printing techniques, nano-inspired biomaterials and enhanced electrochemistry protocols accelerated the possibility of obtaining comparable sensitivity and accuracy, with lower costs and faster protocols [161,183]. Several rapid tests have been proposed in the last decade to detect most relevant viruses, such as HIV, influenza, pneumonia, all with an attempt to face the open challenges both in terms of metrology and low-cost materials to maximize test diffusion [170]. Both amperometric and impedimetric techniques have been recently proposed [184]. The most common detection strategy is based on converting a hybridization event, taking place when the target sequence recognizes its complementarity, into a quantifiable electrical signal [185,186]. Clearly, an amplification of the signal needs to be implemented in order to reach a competitive LOD. Different amplification strategies have been proposed, demonstrating the possibility to reach femto- and atto-molar concentrations of DNA [187,188], thus improving the sensitivity by almost six orders of magnitude compared to standard quantification without amplification [189]. The methods can be grouped into the following categories: (i) enzyme mediated, exploiting the recycle of a single event by the biocatalytic reaction mediated; (ii) nanomaterial-based, exploiting the high surface area of nanoparticles for the high loading of DNA probes [53]; (iii) nucleic acid-based approaches, implementing the local isothermal amplification of the DNA copies before quantification [161]. Of course, to implement most of these portable nucleic acid sensing approaches, the opportunity provided by the printing approach has been exploited, particularly in terms of the integration of the sensing elements with more complex microfluidics polymers or paper [185], [190,191]. This is essential for an accurate quantification of those targets, enabling the possibility to combine sample preparation, purification, amplification and final sensing in the same portable chip [192,193]. Furthermore, the high accuracy obtained in controlling the functionalization of nanostructures and the direct printing of small molecules or specific sequences of DNA and RNA aptamers [194] is opening the door to novel, low-cost, single-use, sensitive tests for very specific applications—for example, single mutation identification [186].

4.3. Proteins

Printed electrochemical biosensors provide novel opportunities for the quantification of proteins, which are peculiar, predictive, diagnostic and prognostic biomarkers of pathophysiological processes [195]. The search for novel protein biomarkers is particularly active for pathologies like cardiovascular disease, cancer or neurodegenerative diseases [196], for which the possibility to rely on novel, low-cost, ultrasensitive, accurate printed biosensors could help to take a step towards early detection, prompting intervention and drug discovery [8,197]. Thus, the early stage protein in particular might be found in blood or in other fluids in concentrations < pg/mL, which are hardly detectable with standard protein analysis [198]. Nowadays, novel printed biosensors have reached LODs of several orders of magnitude lower than the μM range of bulk electrodes, relying on the combination of highly organized novel nanostructures [69,199], magnetic immobilization [200], miniaturized geometries [198], optimized designs to enhance electrochemical parameters and flow control [201,202]. When aiming for large scale and low-cost diagnostic screening, the opportunity for the integration of printed electrochemical sensors with 3D printed modular point-of-care or lab-on-a-chip structures represent a winning strategy [203] with respect to bulky and more expensive optical and mechanical biosensors [133,204]. The techniques adopted for protein quantification using printed electrochemical sensors can mainly be categorized as label free and label based or impedance and voltammetry based. The choice of label-free or label-based techniques strongly depends upon the peculiar characteristics of the protein: molecular weight, electroactivity, surface charge, conformations, trade-off between accuracy, sensitivity and rapidity of analysis required. The impedimetric detection of proteins has been highlighted as a promising label-free tool in recent publications [205,206], combined with novel strategies such as nanostructures or molecular imprinted polymers to increase its sensitivity. Its intrinsic advantage is related to the low complexity of the protocol and the immediate correlation of the protein concentration and impedance value, without additional labels [207]. Voltammetries remain the most promising techniques in terms of the limit of detection and their customization, even if most of the detection protocols, enzyme or label mediated, are still affected by quite a high variability [208]. Alternative interesting detection principles have also been shown in [193] where a preliminary example of inkjet-printed top-gate BioFETs was used for monitoring an immunoreaction by measuring changes in the drain current, paving the way for further use of these types of devices in protein sensing.

4.4. Metabolites and Electrolytes

The development of low-cost, miniaturized, conformable, robust and non-invasive printed electrochemical biosensors for metabolites and electrolytes is attracting more and more interest, in particular with the recent advances in terms of wearable devices and remote sensing [135,209,210]. Thus, since metabolites and electrolytes represent relevant indicators of physio-pathological health found in multiple human fluids (e.g., blood, sweat, saliva), their accurate non-invasive continuous quantification with portable printed devices could serve as a crucial indicator for the prompt detection of a state of alarm, as interestingly highlighted by the most recent research in terms of eHealth and telemedicine [211,212]. The levels of the most common metabolites in human fluids in concentrations usually not lower than μM make them perfect candidates for electrochemical sensing, considering that the LOD of μM can be reached even without nanostructures [88]. However, in order to provide continuous monitoring, peculiar specifications need to be taken in consideration in the design of printed electrochemical biosensors for this class of target analytes: the long-term stability of the materials [213], a proper integration with microfluidic devices to continuously provide the sample to the sensing area [214] and a transduction method compatible with long-term analysis [215].

Regarding metabolites, such as glucose and lactate, the traditional chronoamperometric enzymatic detection techniques have been strongly improved in the last decade thanks to mediators, nanostructures, and a combination of different printable materials and multisensory platform implementation [216]. However, in recent years, several examples of non-enzymatic detection have been proposed [217,218], exploiting the sensitivity of peculiar printable materials and trying to point toward more efficient long-term monitoring. Regarding electrolytes (Na, K, Cl), they represent a key indicator for critical physical and mental health [219]. Their concentration, ranging from µM to mM, can be quantified both in blood and sweat by potentiometric detection due to their intrinsic charges. The state-of-the-art sensing capabilities for potassium, sodium, and pH are ≈10 µA dec^{-1} [220]. Traditionally, using non printed devices, or with commercially available Screen Printed Electrodes (SPEs), the selectivity is ensured by adopting selective membrane. The development of customized Ion Selective Electrodes (ISEs) is particularly focused on novel sensitive and selective printable materials that could substitute the membranes, and improve the electrochemical coupling between the sensing material and target analytes [221].

The active challenges for printed biosensors for metabolites and electrolytes are focused on their effective integration with wearable devices, improving the stability over long periods. In this way, the combination of additive manufacturing with proper microcontrollers and correction algorithms is bringing a real revolution, allowing for the fabrication of a whole new generation of glucose, lactate and electrolyte sensing applications, with embedded electronic and microfluidic control [222].

5. Conclusions

The reviewed research activities spanning across the last two decades in order to highlight how the relevance of electrochemical printed biosensors is widely recognized in fields, including basic research, regenerative medicine, in-hospital analyses and home-based point-of-care. In particular, the possibility of relying on sensitive, robust and low-cost biosensors represents a significant perspective that could create a revolution for the early diagnosis of degenerative and chronic pathologies, in the treatment and control of infectious diseases and in the development of novel solutions for tissue engineering.

Key aspects emerging from our literature analysis highlight the potential that printing technologies can bring to electrochemical biosensors in terms of miniaturization, nano-structuration, novel bio-mimetic materials and non-conventional substrates, as well as integration with microfluidics and embedded electronics. Thus, from a fabrication point of view, promising trends are represented by novel inks and non-conventional substrates for lowering the costs of biosensor fabrication (e.g., paper- or carbon-based materials) [223], by enhanced control of direct surface electrode modifications with nanostructures [98] or binding molecules for the enhancement of sensors' sensitivity and specificity [224]. From a design point of view, promising opportunities are represented by cost-effective realization of fully printed integrated solutions [225], providing customized electronic hardware for ensuring proper biosensors conditioning and wireless data transmission [226], microfluidic circuits to ensure sample preparation, distribution and immobilization and effective strategies for continuous biosensing [214].

Exploiting these opportunities for biosensor fabrication, transduction principles and integration, printing technologies can offer a relevant potential to enlarge the field of application and to face the metrological challenges still affecting biosensing. Thus, an improved signal-to-noise ratio and LODs using nanostructure printing, the reduced cross-correlation by novel printable selective materials, and the increased repeatability and stability achieved with improved curing and printing strategies, represent leading paths that could really help biosensors to make a step towards data validation, robustness and reliability, enabling their commercialization and trusted use by medical personnel and clinical laboratories.

Author Contributions: Conceptualization, S.T., E.S. and M.S.; Writing—Original Draft Preparation, S.T.; Writing—Review & Editing, M.S. and E.S. All authors have read and agreed to the published version of the manuscript.

Funding: This research received no external funding.

Conflicts of Interest: The authors declare no conflict of interest.

References

1. Saengchairat, N.; Tran, T.; Chua, C.-K. A review: Additive manufacturing for active electronic components. *Virtual Phys. Prototyp.* **2016**, *12*, 1–16. [CrossRef]
2. Tan, H.W.; Tran, T.; Chua, C.K. A review of printed passive electronic components through fully additive manufacturing methods. *Virtual Phys. Prototyp.* **2016**, *11*, 271–288. [CrossRef]
3. McEachern, F.; Harvey, E.; Merle, G. Emerging Technologies for the Electrochemical Detection of Bacteria. *Biotechnol. J.* **2020**, *15*, 2000140. [CrossRef] [PubMed]
4. Abdalla, A.; Patel, B.A. 3D-printed electrochemical sensors: A new horizon for measurement of biomolecules. *Curr. Opin. Electrochem.* **2020**, *20*, 78–81. [CrossRef]
5. Munteanu, F.-D.; Titoiu, A.M.; Marty, J.-L.; Vasilescu, A. Detection of antibiotics and evaluation of antibacterial activity with screen-printed electrodes. *Sensors* **2018**, *18*, 901. [CrossRef] [PubMed]
6. Alonso-Lomillo, M.A.; Domínguez-Renedo, O. Screen-printed biosensors in drug analysis. *Curr. Pharm. Anal.* **2017**, *13*, 169–174. [CrossRef]
7. Kozitsina, A.N.; Svalova, T.S.; Malysheva, N.N.; Okhokhonin, A.V.; Vidrevich, M.B.; Brainina, K.Z. Sensors based on bio and biomimetic receptors in medical diagnostic, environment, and food analysis. *Biosensors* **2018**, *8*, 35. [CrossRef]
8. Mincu, N.-B.; Lazar, V.; Stan, D.; Mihailescu, C.M.; Iosub, R.; Mateescu, A.L. Screen-Printed Electrodes (SPE) for in vitro diagnostic purpose. *Diagnostics* **2020**, *10*, 517. [CrossRef]
9. Yáñez-Sedeño, P.; Campuzano, S.; Pingarrón, J.M. Screen-printed electrodes: Promising paper and wearable transducers for (bio)sensing. *Biosensors* **2020**, *10*, 76. [CrossRef]
10. Tuoheti, A.; Aiassa, S.; Criscuolo, F.; Stradolini, F.; Tzouvadaki, I.; Carrara, S.; Demarchi, D. New Approach for Making Standard the Development of Biosensing Devices by a Modular Multi-Purpose Design. *IEEE Trans. Nanobiosci.* **2020**, *19*, 339–346. [CrossRef]
11. Khan, S.; Ali, S.; Bermak, A. Recent developments in printing flexible and wearable sensing electronics for healthcare applications. *Sensors* **2019**, *19*, 1230. [CrossRef] [PubMed]
12. Bhalla, N.; Jolly, P.; Formisano, N.; Estrela, P. Introduction to biosensors. *Essays Biochem.* **2016**, *60*, 1–8.
13. Lavín, Á.; Vicente, J.D.; Holgado, M.; Laguna, M.F.; Casquel, R.; Santamaria, B.; Maigler, M.V.; Hernandez, A.L.; Ramirez, Y. On the Determination of Uncertainty and Limit of Detection in Label-Free Biosensors. *Sensors* **2018**, *18*, 2038. [CrossRef]
14. Armbruster, D.A.; Pry, T. Limit of blank, limit of detection and limit of quantitation. *Clin. Biochem. Rev.* **2008**, *29* (Suppl. 1), S49–S52.
15. Kokkinos, C.; Economou, A. Recent advances in voltammetric, amperometric and ion-selective (bio)sensors fabricated by microengineering manufacturing approaches. *Curr. Opin. Electrochem.* **2020**, *23*, 21–25. [CrossRef]
16. Najeeb, M.A.; Ahmad, Z.; Shakoor, R.A.; Mohamed, A.M.A.; Kahraman, R. A novel classification of prostate specific antigen (PSA) biosensors based on transducing elements. *Talanta* **2017**, *168*, 52–61. [CrossRef] [PubMed]
17. Afzal, A.; Mujahid, A.; Schirhagl, R.; Bajwa, S.Z.; Latif, U.; Feroz, S. Gravimetric viral diagnostics: QCM based biosensors for early detection of viruses. *Chemosensors* **2017**, *5*, 7. [CrossRef]
18. Nolan, P.; Auer, S.; Spehar, A.; Oplatowska-Stachowiak, M.; Campbell, K. Evaluation of Mass Sensitive Micro-Array biosensors for their feasibility in multiplex detection of low molecular weight toxins using mycotoxins as model compounds. *Talanta* **2020**, *222*, 12152.
19. Arlett, J.; Myers, E.B.; Roukes, M. Comparative Advantages of Mechanical Biosensors. *Nat. Nanotechnol.* **2011**, *6*, 203–215. [CrossRef]
20. Rezabakhsh, A.; Rahbarghazi, R.; Fathi, F. Surface plasmon resonance biosensors for detection of Alzheimer's biomarkers; an effective step in early and accurate diagnosis. *Biosens. Bioelectron.* **2020**, *167*, 112511. [CrossRef]
21. Sharma, S.; Kumari, R.; Varshney, S.K.; Lahiri, B. Optical biosensing with electromagnetic nanostructures. *Rev. Phys.* **2020**, *5*, 100044. [CrossRef]
22. Tabassum, S.; Kumar, R. Advances in Fiber-Optic Technology for Point-of-Care Diagnosis and In Vivo Biosensing. *Adv. Mater. Technol.* **2020**, *5*, 19000792. [CrossRef]
23. Méjard, R.; Griesser, H.J.; Thierry, B. Optical biosensing for label-free cellular studies. *TrAC—Trends Anal. Chem.* **2014**, *53*, 178–186. [CrossRef]

24. Lucarelli, F.; Tombelli, S.; Minunni, M.; Marrazza, G.; Mascini, M. Electrochemical and piezoelectric DNA biosensors for hybridisation detection. *Anal. Chim. Acta* **2008**, *609*, 139–159. [CrossRef]
25. Muñoz, J.; Pumera, M. 3D-printed biosensors for electrochemical and optical applications. *TrAC Trends Anal. Chem.* **2020**, *128*, 115933. [CrossRef]
26. Yu, H.L.L.; Maslova, A.; Hsing, I.-M. Rational Design of Electrochemical DNA Biosensors for Point-of-Care Applications. *Chem. Electro. Chem.* **2017**, *4*, 795–805. [CrossRef]
27. Menon, S.; Mathew, M.R.; Sam, S.; Keerthi, K.; Kumar, K.G. Recent advances and challenges in electrochemical biosensors for emerging and re-emerging infectious diseases. *J. Electroanal. Chem.* **2020**, *878*, 114596. [CrossRef]
28. Han, Y.; Dong, J. Electrohydrodynamic printing for advanced micro/nanomanufacturing: Current progresses, opportunities, and challenges. *J. Micro Nano-Manuf.* **2018**. [CrossRef]
29. Kamanina, O.A.; Kamanin, S.S.; Kharkova, A.S.; Arlyapov, V.A. Glucose biosensor based on screen-printed electrode modified with silicone sol–gel conducting matrix containing carbon nanotubes. *3 Biotech.* **2019**, *9*, 290. [CrossRef]
30. Soni, D.; Ahmad, R.; Dubey, S. Biosensor for the detection of Listeria monocytogenes: Emerging trends. *Crit. Rev. Microbiol.* **2018**, *44*, 590–608. [CrossRef]
31. Brett, A.M.O.; Serrano, S.H.P.; Gutz, I.G.R.; La-Scalea, M.A. Comparison of the Voltammetric Behavior of Metronidazole at a DNA-Modified Glassy Carbon Electrode, a Mercury Thin Film Electrode and a Glassy Carbon Electrode. *Electroanalysis* **1997**, *9*, 110–114. [CrossRef]
32. Nesaei, S.; Song, Y.; Wang, Y.; Ruan, X.; Du, D.; Gozen, A.; Lin, Y. Micro additive manufacturing of glucose biosensors: A feasibility study. *Anal. Chim. Acta* **2018**, *1043*, 142–149. [CrossRef] [PubMed]
33. Hashim, U.; Salleh, S.; Rahman, S.F.A.; Abdullah, A.R.A.J. Design and fabrication of Nanowire-based conductance biosensor using spacer patterning technique. In Proceedings of the 2008 International Conference on Electronic Design, ICED 2008, Penang, Malaysia, 1–3 December 2008; IEEE: New York, NY, USA, 2008.
34. Sokolov, A.N.; Roberts, M.E.; Bao, Z. Fabrication of low-cost electronic biosensors. *Mater. Today* **2009**, *12*, 12–20. [CrossRef]
35. Raymundo-Pereira, P.A.; Baccarin, M.; Oliveira, O.N., Jr.; Janegitz, B.C. Thin Films and Composites Based on Graphene for Electrochemical Detection of Biologically-relevant Molecules. *Electroanalysis* **2018**, *30*, 1888–1896. [CrossRef]
36. Cotte, S.; Baraket, A.; Bessueille, F.; Gout, S.; Yaakoubi, N.; Leonard, D.; Errachid, A. Fabrication of Microelectrodes Using Original 'Soft Lithography' Processes. In *New Sensors and Processing Chain*; Wiley Online Library: Hoboken, NJ, USA, 2014; Volume 9781848216266, pp. 1–9.
37. Tran, K.T.M.; Nguyen, T.D. Lithography-based methods to manufacture biomaterials at small scales. *J. Sci. Adv. Mater. Devices* **2017**, *2*, 1–14. [CrossRef]
38. Castrovilli, M.C.; Bolognesi, P.; Chiarinelli, J.; Avaldi, L.; Cartoni, A.; Calandra, P.; Tempesta, E.; Giardi, M.T.; Antonacci, A.; Arduini, F.; et al. Electrospray deposition as a smart technique for laccase immobilisation on carbon black-nanomodified screen-printed electrodes. *Biosens. Bioelectron.* **2020**, *163*, 112299. [CrossRef]
39. Al-Dhahebi, A.M.; Gopinath, S.C.B.; Saheed, M.S.M. Graphene impregnated electrospun nanofiber sensing materials: A comprehensive overview on bridging laboratory set-up to industry. *Nano Converg.* **2020**, *7*, 1–23. [CrossRef]
40. Liu, Y.; Hao, M.; Chen, Z.; Liu, L.; Liu, Y.; Yang, W.; Ramakrishna, S. A review on recent advances in application of electrospun nanofiber materials as biosensors. *Curr. Opin. Biomed. Eng.* **2020**, *13*, 174–189. [CrossRef]
41. Willmann, J.; Stocker, D.; Dörsam, E. Characteristics and evaluation criteria of substrate-based manufacturing. Is roll-to-roll the best solution for printed electronics? *Org. Electron.* **2014**, *15*, 1631–1640. [CrossRef]
42. Tonello, S.; Serpelloni, M.; Lopomo, N.F.; Abate, G.; Uberti, D.L.; Sardini, E. Screen-Printed Biosensors for the Early Detection of Biomarkers Related to Alzheimer Disease: Preliminary Results. In *Procedia Engineering*; Elsevier: New York, NY, USA, 2016; Volume 168.
43. Lau, G.-K.; Shrestha, M. Ink-Jet Printing of Micro-Electro-Mechanical Systems (MEMS). *Micromachines* **2017**, *8*, 194. [CrossRef]
44. Agarwala, S.; Goh, G.L.; Yeong, W.Y. Optimizing aerosol jet printing process of silver ink for printed electronics. *IOP Conf. Ser. Mater. Sci. Eng.* **2017**, *191*, 12027. [CrossRef]

45. Khan, S.; Lorenzelli, L.; Dahiya, R.S. Technologies for printing sensors and electronics over large flexible substrates: A review. *IEEE Sens. J.* **2015**, *15*, 3164–3185. [CrossRef]
46. Mondal, K.; McMurtrey, M.D. Present status of the functional advanced micro-, nano-printings—A mini review. *Mater. Today Chem.* **2020**, *17*. [CrossRef]
47. Grünwald, S. Reproducible dispensing of liquids in the nanolitre range. *Adhes. Adhes. Sealants* **2018**, *15*, 28–31. [CrossRef]
48. Abas, M.; Salman, Q.; Khan, A.M.; Rahman, K. Direct ink writing of flexible electronic circuits and their characterization. *J. Brazilian Soc. Mech. Sci. Eng.* **2019**, *41*, 563. [CrossRef]
49. Yang, H.; Rahman, T.; Du, D.; Panat, R.; Lin, Y. 3-D Printed Adjustable Microelectrode Arrays for Electrochemical Sensing and Biosensing. *Sens. Actuators. B. Chem.* **2016**, *230*, 600–606. [CrossRef]
50. Hoffman, J.; Hwang, S.; Ortega, A.; Kim, N.-S.; Moon, K.-S. The standardization of printable materials and direct writing systems. *J. Electron. Packag. Trans. ASME* **2013**, *135*. [CrossRef]
51. Ramasamy, M.; Varadan, V.K. 3D printing of nano-and micro-structures. In *Proceedings of SPIE—The International Society for Optical Engineering*; International Society for Optics and Photonics: Hague, The Netherlands, 2016; Volume 9802.
52. Dziąbowska, K.; Czaczyk, E.; Nidzworski, D. Application of Electrochemical Methods in Biosensing Technologies. *Biosens. Technol. Detect. Pathog. A Prosp. Way Rapid Anal.* **2018**. [CrossRef]
53. Nagar, B.; Balsells, M.; de la Escosura-Muñiz, A.; Gomez-Romero, P.; Merkoçi, A. Fully printed one-step biosensing device using graphene/AuNPs composite. *Biosens. Bioelectron.* **2019**, *129*, 238–244. [CrossRef] [PubMed]
54. Wang, Y.; Ye, Z.; Ying, Y. New trends in impedimetric biosensors for the detection of foodborne pathogenic bacteria. *Sensors* **2012**, *12*, 3449–3471. [CrossRef]
55. Soleymani, L.; Li, F. Mechanistic Challenges and Advantages of Biosensor Miniaturization into the Nanoscale. *ACS Sensors* **2017**, *2*, 458–467. [CrossRef] [PubMed]
56. Zhang, W.; Wang, R.; Luo, F.; Wang, P.; Lin, Z. Miniaturized electrochemical sensors and their point-of-care applications. *Chin. Chem. Lett.* **2020**, *31*, 589–600. [CrossRef]
57. Manjakkal, L.; Shakthivel, D.; Dahiya, R. Flexible Printed Reference Electrodes for Electrochemical Applications. *Adv. Mater. Technol.* **2018**, *3*, 1800252. [CrossRef]
58. Søpstad, S.; Johannessen, E.A.; Imenes, K. Analytical errors in biosensors employing combined counter/pseudo-reference electrodes. *Results Chem.* **2020**, *2*, 100028. [CrossRef]
59. Faria, A.M.; Peixoto, E.B.M.I.; Adamo, C.B.; Flacker, A.; Longo, E.; Mazon, T. Controlling parameters and characteristics of electrochemical biosensors for enhanced detection of 8-hydroxy-2′-deoxyguanosine. *Sci. Rep.* **2019**, *9*, 7411. [CrossRef] [PubMed]
60. Sopstad, S.; Imenes, K.; Johannessen, E.A. Chloride and pH Determination on a Wireless, Flexible Electrochemical Sensor Platform. *IEEE Sens. J.* **2020**, *20*, 599–609. [CrossRef]
61. Diamond, D. Analytical electrochemistry—Analytical Electrochemistry, by Joseph Wang, VCH, Weinheim, 1994, xii + 198 pages, DM 98.00, ISBN 1-56081-572-2. *Trends Anal. Chem.* **1996**, *15*, X–XI. [CrossRef]
62. Thapliyal, N.; Chiwunze, T.; Karpoormath, R.; Goyal, R.; Patel, H.; Srinivasulu, C. Research Progress in Electroanalytical Techniques for Determination of Antimalarial Drugs in Pharmaceutical and Biological Samples. *RSC Adv.* **2016**, *6*, 57580–57602. [CrossRef]
63. Gwon, K.; Lee, S.; Nam, H.; Shin, J.H. Disposable strip-type biosensors for amperometric determination of galactose. *J. Electrochem. Sci. Technol.* **2020**, *11*, 310–317. [CrossRef]
64. Leva-Bueno, J.; Peyman, S.A.; Millner, P.A. A review on impedimetric immunosensors for pathogen and biomarker detection. *Med. Microbiol. Immunol.* **2020**, *209*, 343–362. [CrossRef]
65. Zehani, N.; Dzyadevych, S.; Kherrat, R.; Jaffrezic-Renault, N. Sensitive impedimetric biosensor for direct detection of diazinon based on lipases. *Front. Chem.* **2014**, *2*, 44. [CrossRef]
66. Ariffin, E.Y.; Heng, L.Y.; Tan, L.L.; Karim, N.H.A.; Hasbullah, S.A. A highly sensitive impedimetric DNA biosensor based on hollow silica microspheres for label-free determination of E. Coli. *Sensors* **2020**, *20*, 1279. [CrossRef]
67. Ding, J.; Qin, W. Recent advances in potentiometric biosensors. *TrAC Trends Anal. Chem.* **2020**, *124*, 115803. [CrossRef]
68. Jaffrezic-Renault, N.; Dzyadevych, S.V. Conductometric Microbiosensors for Environmental Monitoring. *Sensors* **2008**, *8*, 2569–2588. [CrossRef] [PubMed]

69. Hammond, J.L.; Formisano, N.; Estrela, P.; Carrara, S.; Tkac, J. Electrochemical biosensors and nanobiosensors. *Essays Biochem.* **2016**, *60*, 69–80.
70. Rocchitta, G.; Spanu, A.; Babudieri, S.; Latte, G.; Madeddu, G.; Galleri, G.; Nuvoli, S.; Bagella, P.; Demartis, M.; Fiore, V.; et al. Analytical Problems in Exposing Amperometric Enzyme Biosensors to Biological Fluids. *Sensors* **2016**, *16*, 780. [CrossRef]
71. Pemberton, R.M.; Xu, J.; Pittson, R.; Drago, G.A.; Griffiths, J.; Jackson, S.K.; Hart, J.P. A screen-printed microband glucose biosensor system for real-time monitoring of toxicity in cell culture. *Biosens. Bioelectron.* **2011**, *26*, 2448–2453. [CrossRef]
72. Mistry, K.K.; Layek, K.; Mahapatra, A.; RoyChaudhuri, C.; Saha, H. A review on amperometric-type immunosensors based on screen-printed electrodes. *Analyst* **2014**, *139*, 2289–2311. [CrossRef]
73. Alarcon-Angeles, G.; Álvarez-Romero, G.A.; Merkoçi, A. Electrochemical biosensors: Enzyme kinetics and role of nanomaterials. In *Encyclopedia of Interfacial Chemistry: Surface Science and Electrochemistry*; Elsevier: Amsterdam, The Netherlands, 2018; pp. 140–155. [CrossRef]
74. Pal, K.; Kraatz, H.-B.; Khasnobish, A.; Bag, S.; Banerjee, I.; Kuruganti, U. *Bioelectronics and Medical Devices: From Materials to Devices—Fabrication, Applications and Reliability*; Elsevier: Amsterdam, The Netherlands, 2019. [CrossRef]
75. Li, H.; Liu, X.; Li, L.; Mu, X.; Genov, R.; Mason, A.J. CMOS Electrochemical Instrumentation for Biosensor Microsystems: A Review. *Sensors* **2016**, *17*, 74. [CrossRef]
76. Bahadır, E.B.; Sezgintürk, M.K. A review on impedimetric biosensors. *Artif. Cells Nanomed. Biotechnol.* **2016**, *44*, 248–262. [CrossRef]
77. Hopkins, J.; Fidanovski, K.; Lauto, A.; Mawad, D. All-Organic Semiconductors for Electrochemical Biosensors: An Overview of Recent Progress in Material Design. *Front. Bioeng. Biotechnol.* **2019**, *7*, 237. [CrossRef]
78. Bogomolova, A.; Komarova, E.; Reber, K.; Gerasimov, T.; Yavuz, O.; Bhatt, S.; Aldissi, M. Challenges of Electrochemical Impedance Spectroscopy in Protein Biosensing. *Anal. Chem.* **2009**, *81*, 3944–3949. [CrossRef]
79. Carminati, M.; Ferrari, G.; Bianchi, D.; Sampietro, M. *Impedance Spectroscopy for Biosensing: Circuits and Applications*; Springer: New York, NY, USA, 2015; pp. 1–24.
80. Tonello, S.; Bianchetti, A.; Braga, S.; Almici, C.; Marini, M.; Piovani, G.; Guindani, M.; Dey, K.; Sartore, L.; Re, F.; et al. Impedance-based monitoring of mesenchymal stromal cell three-dimensional proliferation using aerosol jet printed sensors: A tissue engineering application. *Materials* **2020**, *13*, 2231. [CrossRef]
81. Aggas, J.R.; Abasi, S.; Phipps, J.F.; Podstawczyk, D.A.; Guiseppi-Elie, A. Microfabricated and 3-D printed electroconductive hydrogels of PEDOT:PSS and their application in bioelectronics. *Biosens. Bioelectron.* **2020**, *168*, 112568. [CrossRef] [PubMed]
82. Li, J.; Qin, W. An integrated all-solid-state screen-printed potentiometric sensor based on a three-dimensional self-assembled graphene aerogel. *Microchem. J.* **2020**, *159*, 105453. [CrossRef]
83. Tehrani, Z.; Whelan, S.P.; Mostert, A.B.; Paulin, J.V.; Ali, M.M.; Ahmadi, E.D.; Graeff, C.F.O.; Guy, O.J.; Gethin, D.T. Printable and flexible graphene pH sensors utilising thin film melanin for physiological applications. *2D Mater.* **2020**, *7*, 24008. [CrossRef]
84. Mishra, R.K.; Barfidokht, A.; Karajic, A.; Sempionatto, J.R.; Wang, J.; Wang, J. Wearable potentiometric tattoo biosensor for on-body detection of G-type nerve agents simulants. *Sens. Actuators B Chem.* **2018**, *273*, 966–972. [CrossRef]
85. Gauglitz, G. Analytical evaluation of sensor measurements. *Anal. Bioanal. Chem.* **2018**, *410*, 5–13. [CrossRef]
86. Mendes, P.M. Cellular nanotechnology: Making biological interfaces smarter. *Chem. Soc. Rev.* **2013**, *42*, 9207–9218. [CrossRef]
87. Chen, M.; Lee, H.; Yang, J.; Xu, Z.; Huang, N.; Chan, B.P.; Kim, J.T. Parallel, Multi-Material Electrohydrodynamic 3D Nanoprinting. *Small* **2020**, *16*, 1906402. [CrossRef] [PubMed]
88. Carrara, S.; Baj-Rossi, C.; Boero, C.; de Micheli, G. Do carbon nanotubes contribute to electrochemical biosensing? *Electrochim. Acta* **2014**, *128*, 102–112. [CrossRef]
89. Hail, C.U.; Höller, C.; Matsuzaki, K.; Rohner, P.; Renger, J.; Sandoghdar, V.; Poulikakos, D.; Eghlidi, H. Nanoprinting organic molecules at the quantum level. *Nat. Commun.* **2019**, *10*, 1880. [CrossRef]
90. Al-Milaji, K.N.; Secondo, R.R.; Ng, T.N.; Kinsey, N.; Zhao, H. Interfacial Self-Assembly of Colloidal Nanoparticles in Dual-Droplet Inkjet Printing. *Adv. Mater. Interfaces* **2018**, *5*, 1701561. [CrossRef]
91. Zhu, C.; Yang, G.; Li, H.; Du, D.; Lin, Y. Electrochemical sensors and biosensors based on nanomaterials and nanostructures. *Anal. Chem.* **2015**, *87*, 230–249. [CrossRef]

92. Masud, M.K.; Mahmudunnabi, R.G.; Aziz, N.B.; Stevens, C.H.; Do-Ha, D.; Yang, S.; Blair, I.P.; Hossain, M.S.A.; Shim, Y.; Ooi, L.; et al. Sensitive Detection of Motor Neuron Disease Derived Exosomal miRNA Using Electrocatalytic Activity of Gold-Loaded Superparamagnetic Ferric Oxide Nanocubes. *Chem. Electr. Chem.* **2020**, *7*, 3459–3467.
93. Nie, C.; Ma, L.; Li, S.; Fan, X.; Yang, Y.; Cheng, C.; Zhao, W.; Zhao, C. Recent progresses in graphene based bio-functional nanostructures for advanced biological and cellular interfaces. *Nano Today* **2019**, *26*, 57–97. [CrossRef]
94. Das, S.R.; Nian, Q.; Cargill, A.A.; Hondred, J.A.; Ding, S.; Saei, M.; Cheng, G.J.; Claussen, J.C. 3D nanostructured inkjet printed graphene: Via UV-pulsed laser irradiation enables paper-based electronics and electrochemical devices. *Nanoscale* **2016**, *8*, 15870–15879. [CrossRef]
95. Sarah, D.U.; Tonello, M.M.; Carrara, E.S.S.; Lopomo, N.F.; Serpelloni, M. Enhanced Sensing of Interleukin 8 by Stripping Voltammetry: Carbon Nanotubes versus Fullerene. In *EMBEC NBC 2017-Joint Conference European Medical Biology Engineering Conference Nordic-Baltic Conference Biomedical Engineering Medical Physical*; Springer: Cham, Switzerland, 2018; pp. 213–218.
96. Tonello, S.; Stradolini, F.; Abate, G.; Uberti, D.; Serpelloni, M.; Carrara, S.; Sardini, E. Electrochemical detection of different p53 conformations by using nanostructured surfaces. *Sci. Rep.* **2019**, *9*, 17347. [CrossRef]
97. Heller, I.; Männik, J.; Lemay, S.G.; Dekker, C. Optimizing the Signal-to-Noise Ratio for Biosensing with Carbon Nanotube Transistors. *Nano Lett.* **2009**, *9*, 377–382. [CrossRef]
98. Wongkaew, N.; Simsek, M.; Griesche, C.; Baeumner, A.J. Functional Nanomaterials and Nanostructures Enhancing Electrochemical Biosensors and Lab-on-a-Chip Performances: Recent Progress, Applications, and Future Perspective. *Chem. Rev.* **2019**, *119*, 120–194. [CrossRef]
99. Cantù, E.; Tonello, S.; Abate, G.; Uberti, D.; Sardini, E.; Serpelloni, M. Aerosol Jet Printed 3D Electrochemical Sensors for Protein Detection. *Sensors* **2018**, *18*, 3719. [CrossRef]
100. Hainaut, P.; Mann, K. Zinc binding and redox control of p53 structure and function. *Antioxid. Redox Signal.* **2001**, *3*, 611–623. [CrossRef]
101. Sanati, A.; Jalali, M.; Raeissi, K.; Karimzadeh, F.; Kharaziha, M.; Mahshid, S.S.; Mahshid, S. A review on recent advancements in electrochemical biosensing using carbonaceous nanomaterials. *Microchim. Acta* **2019**, *186*, 773. [CrossRef]
102. Lago, N.; Buonomo, M.; Imran, S.; Bertani, R.; Wrachien, N.; Bortolozzi, M.; Pedersen, M.G.; Cester, A. TIPS-Pentacene as Biocompatible Material for Solution Processed High-Performance Electronics Operating in Water. *IEEE Electron Device Lett.* **2018**, *39*, 1401–1404. [CrossRef]
103. Riera-Galindo, S.; Leonardi, F.; Pfattner, R.; Mas-Torrent, M. Organic Semiconductor/Polymer Blend Films for Organic Field-Effect Transistors. *Adv. Mater. Technol.* **2019**, *4*, 19000104. [CrossRef]
104. Liu, Y.; Turner, A.P.F.; Zhao, M.; Mak, W.C. Processable enzyme-hybrid conductive polymer composites for electrochemical biosensing. *Biosens. Bioelectron.* **2018**, *100*, 374–381. [CrossRef]
105. Stříteský, S.; Markova, A.; Vitevcek, J.; Vsafavrikova, E.; Hrabal, M.; Kubavc, L.; Kubala, L.; Weiter, M.; Vala, M. Printing inks of electroactive polymer PEDOT:PSS: The study of biocompatibility, stability, and electrical properties. *J. Biomed. Mater. Res. A* **2018**, *106*, 1121–1128. [CrossRef]
106. Sessolo, M.; Khodagholy, D.; Rivnay, J.; Maddalena, F.; Gleyzes, M.; Steidl, E.; Buisson, B.; Malliaras, G.G. Easy-to-fabricate conducting polymer microelectrode arrays. *Adv. Mater.* **2013**, *25*, 2135–2139. [CrossRef] [PubMed]
107. Wang, J.; Ye, D.; Meng, Q.; Di, C.; Zhu, D. Advances in Organic Transistor-Based Biosensors. *Adv. Mater. Technol.* **2020**, *5*, 2000218. [CrossRef]
108. Lin, P.; Yan, F. Organic thin-film transistors for chemical and biological sensing. *Adv. Mater.* **2012**, *24*, 34–51. [CrossRef]
109. Khan, M.A.; Cantù, E.; Tonello, S.; Serpelloni, M.; Lopomo, N.F.; Sardini, E. A Review on Biomaterials for 3D Conductive Scaffolds for Stimulating and Monitoring Cellular Activities. *Appl. Sci.* **2019**, *9*, 961. [CrossRef]
110. Wang, L.; Lou, Z.; Wang, K.; Zhao, S.; Yu, P.; Wei, W.; Wang, D.; Han, W.; Jiang, K.; Shen, G.; et al. Biocompatible and Biodegradable Functional Polysaccharides for Flexible Humidity Sensors. *Research* **2020**, *2020*, 8716847. [CrossRef]
111. Alsuradi, H.; Yoo, J. Screen Printed Passives and Interconnects on Bio-Degradable Medical Hydrocolloid Dressing for Wearable Sensors. *Sci. Rep.* **2019**, *9*, 17467. [CrossRef]

112. Cosseddu, P.; Basirico, L.; Loi, A.; Lai, S.; Maiolino, P.; Baglini, E.; Denei, S.; Mastrogiovanni, F.; Cannata, G.; Bonfiglio, A. Inkjet printed Organic Thin Film Transistors based tactile transducers for artificial robotic skin. In Proceedings of the IEEE RAS and EMBS International Conference on Biomedical Robotics and Biomechatronics, Rome, Italy, 24–27 June 2012. [CrossRef]
113. Di Novo, N.G.; Cantù, E.; Tonello, S.; Sardini, E.; Serpelloni, M. Support-Material-Free Microfluidics on an Electrochemical Sensors Platform by Aerosol Jet Printing. *Sensors* **2019**, *19*, 1842. [CrossRef]
114. Serien, D.; Sugioka, K. Three-Dimensional Printing of Pure Proteinaceous Microstructures by Femtosecond Laser Multiphoton Cross-Linking. *ACS Biomater. Sci. Eng.* **2020**, *6*, 1279–1287. [CrossRef]
115. Diogo, G.S.; Marques, C.F.; Sotelo, C.G.; Pérez-Martín, R.I.; Pirraco, R.P.; Reis, R.L.; Silva, T.H. Cell-Laden Biomimetically Mineralized Shark-Skin-Collagen-Based 3D Printed Hydrogels for the Engineering of Hard Tissues. *ACS Biomater. Sci. Eng.* **2020**, *6*, 3664–3672. [CrossRef]
116. Marziano, M.; Tonello, S.; Cantu, E.; Abate, G.; Vezzoli, M.; Rungratanawanich, W.; Serpelloni, M.; Lopomo, N.F.; Memo, M.; Sardini, E.; et al. Monitoring Caco-2 to enterocyte-like cells differentiation by means of electric impedance analysis on printed sensors. *Biochim. Biophys. Acta Gen. Subj.* **2019**, *1863*, 893–902. [CrossRef]
117. Nam, Y.; Wheeler, B.C. In vitro microelectrode array technology and neural recordings. *Crit. Rev. Biomed. Eng.* **2011**, *39*, 45–61.
118. Shen, J.; Dudik, L.; Liu, C.C. An iridium nanoparticles dispersed carbon based thick film electrochemical biosensor and its application for a single use, disposable glucose biosensor. *Sensors Actuators B Chem.* **2007**, *125*, 106–113. [CrossRef]
119. Gao, A.; Zou, N.; Dai, P.; Lu, N.; Li, T.; Wang, Y.; Zhao, J.; Mao, H. Signal-to-Noise Ratio Enhancement of Silicon Nanowires Biosensor with Rolling Circle Amplification. *Nano Lett.* **2013**, *13*, 4123–4130. [CrossRef]
120. Tonello, S.; Borghetti, M.; Lopomo, N.F.; Serpelloni, M.; Sardini, E.; Marziano, M.; Serzanti, M.; Uberti, D.; Dell'era, P.; Inverardi, N.; et al. Ink-jet printed stretchable sensors for cell monitoring under mechanical stimuli: A feasibility study. *J. Mech. Med. Biol.* **2019**, 19. [CrossRef]
121. Lee, C.H.; Kim, D.R.; Zheng, X. Fabrication of Nanowire Electronics on Nonconventional Substrates by Water-Assisted Transfer Printing Method. *Nano Lett.* **2011**, *11*, 3435–3439. [CrossRef]
122. Shafiee, H.; Asghar, W.; Inci, F.; Yuksekkaya, M.; Jahangir, M.; Zhang, M.H.; Durmus, N.G.; Gurkan, U.A.; Kuritzkes, D.R.; Demirci, U.; et al. Paper and Flexible Substrates as Materials for Biosensing Platforms to Detect Multiple Biotargets. *Sci. Rep.* **2015**, *5*, 8719. [CrossRef]
123. Baby, T.T.; Marques, G.C.; Neuper, F.; Singaraju, S.A.; Garlapati, S.; von Seggern, F.; Kruk, R.; Dasgupta, S.; Sykora, B.; Breitung, B.; et al. Printing Technologies for Integration of Electronic Devices and Sensors. In *NATO Science for Peace and Security Series C: Environmental Security*; Springer: Cham, Switzerland, 2020; pp. 1–34.
124. Kuswandi, B.; Ensafi, A.A. Perspective—Paper-Based Biosensors: Trending Topic in Clinical Diagnostics Developments and Commercialization. *J. Electrochem. Soc.* **2020**, *167*, 37509. [CrossRef]
125. Tortorich, R.P.; Shamkhalichenar, H.; Choi, J.-W. Inkjet-Printed and Paper-Based Electrochemical Sensors. *Appl. Sci.* **2018**, *8*, 288. [CrossRef]
126. Lehmhus, D.; Aumund-Kopp, C.; Petzoldt, F.; Godlinski, D.; Haberkorn, A.; Zollmer, V.; Busse, M. Customized Smartness: A Survey on Links between Additive Manufacturing and Sensor Integration. *Procedia Technol.* **2016**, *26*, 284–301. [CrossRef]
127. Lu, B.-H.; Lan, H.-B.; Liu, H.-Z. Additive manufacturing frontier: 3D printing electronics. *Opto-Electron. Adv.* **2018**, *1*, 17000401–17000410. [CrossRef]
128. Shkodra, B.; Abera, B.D.; Cantarella, G.; Douaki, A.; Avancini, E.; Petti, L.; Lugli, P. Flexible and Printed Electrochemical Immunosensor Coated with Oxygen Plasma Treated SWCNTs for Histamine Detection. *Biosensors* **2020**, *10*, 35. [CrossRef]
129. Yang, X.; Cheng, H. Recent developments of flexible and stretchable electrochemical biosensors. *Micromachines* **2020**, *11*, 243. [CrossRef] [PubMed]
130. Ermis, M.; Antmen, E.; Hasirci, V. Micro and Nanofabrication methods to control cell-substrate interactions and cell behavior: A review from the tissue engineering perspective. *Bioact. Mater.* **2018**, *3*, 355–369. [CrossRef]

131. Didier, C.; Kundu, A.; Rajaraman, S. Capabilities and limitations of 3D printed microserpentines and integrated 3D electrodes for stretchable and conformable biosensor applications. *Microsyst. Nanoeng.* **2020**, *6*, 15. [CrossRef]
132. Linghu, C.; Zhang, S.; Wang, C.; Song, J. Transfer printing techniques for flexible and stretchable inorganic electronics. *npj Flex. Electron.* **2018**, *2*, 26. [CrossRef]
133. Solhi, E.; Hasanzadeh, M.; Babaie, P. Electrochemical paper-based analytical devices (ePADs) toward biosensing: Recent advances and challenges in bioanalysis. *Anal. Methods* **2020**, *12*, 1398–1414. [CrossRef]
134. Punjiya, M.; Moon, C.H.; Matharu, Z.; Nejad, H.R.; Sonkusale, S. A three-dimensional electrochemical paper-based analytical device for low-cost diagnostics. *Analyst* **2018**, *143*, 1059–1064. [CrossRef]
135. Cinti, S.; Moscone, D.; Arduini, F. Preparation of paper-based devices for reagentless electrochemical (bio)sensor strips. *Nat. Protoc.* **2019**, *14*, 2437–2451. [CrossRef] [PubMed]
136. Kalambate, P.K.; Rao, Z.; Wu, J.; Shen, Y.; Boddula, R.; Huang, Y. Electrochemical (bio) sensors go green. *Biosens. Bioelectron.* **2020**, *163*, 112270. [CrossRef]
137. Ratajczak, K.; Stobiecka, M. High-performance modified cellulose paper-based biosensors for medical diagnostics and early cancer screening: A concise review. *Carbohydr. Polym.* **2020**, *229*, 115463. [CrossRef]
138. Cooper, C.; Hughes, B. Aerosol Jet Printing of Electronics: An Enabling Technology for Wearable Devices. In *2020 Pan Pacific Microelectronics Symposium, Pan Pacific 2020*; IEEE: New York, NY, USA, 2020.
139. Cantù, E.; Soprani, M.; Ponzoni, A.; Sardini, E.; Serpelloni, M. Preliminary analysis on cellulose-based gas sensor by means of aerosol jet printing and photonic sintering. In the BIODEVICES 2020—13th International Conference on Biomedical Electronics and Devices, Proceedings; Part of 13th International Joint Conference on Biomedical Engineering Systems and Technologies, BIOSTEC 2020, Valletta, Malta, 24–26 February 2020; Multidisciplinary Digital Publishing Institute: Basel, Switzerland, 2020; pp. 200–206.
140. Serpelloni, M.; Cantù, E.; Borghetti, M.; Sardini, E. Printed smart devices on cellulose-based materials by means of aerosol-jet printing and photonic curing. *Sensors* **2020**, *20*, 841. [CrossRef]
141. Smith, M.; Choi, Y.; Boughey, C.; Kar-Narayan, S. Controlling and assessing the quality of aerosol jet printed features for large area and flexible electronics. *Flex. Print. Electron.* **2017**, *2*, 15004. [CrossRef]
142. Castillo-León, J.; Svendsen, W.E. *Lab-on-a-Chip Devices and Micro-Total Analysis Systems: A Practical Guide*; Springer International Publishing: Cham, Switzerland, 2014.
143. Lu, J.; Ge, S.; Ge, L.; Yan, M.; Yu, J. Electrochemical DNA sensor based on three-dimensional folding paper device for specific and sensitive point-of-care testing. *Electrochim. Acta* **2012**, *80*, 334–341. [CrossRef]
144. Dong, Z.; Ma, J.; Jiang, L. Manipulating and Dispensing Micro/Nanoliter Droplets by Superhydrophobic Needle Nozzles. *ACS Nano* **2013**, *7*, 10371–10379. [CrossRef]
145. Sharafeldin, M.; Jones, A.; Rusling, J.F. 3D-Printed Biosensor Arrays for Medical Diagnostics. *Micromachines* **2018**, *9*, 394. [CrossRef]
146. Khurana, J.B.; Dinda, S.; Simpson, T.W. Active—Z printing: A new approach to increasing 3D printed part strength. In *Solid Freeform Fabrication 2017: Proceedings of the 28th Annual International Solid Freeform Fabrication Symposium—An Additive Manufacturing Conference, SFF 2017, Austin, TX, USA, 7–9 August 2017*; pp. 1627–1644.
147. Lee, K.; Yoon, T.; Yang, H.; Cha, S.; Cheon, Y.; Kashefi-Kheyrabadi, L.; Jung, H. All-in-one platform for salivary cotinine detection integrated with a microfluidic channel and an electrochemical biosensor. *Lab Chip* **2020**, *20*, 320–331. [CrossRef] [PubMed]
148. Bodini, A.; Cantu', E.; Serpelloni, M.; Sardini, E.; Tonello, S. Design and implementation of a microsensor platform for protein detection realized via 3-D printing. In *2018 IEEE Sensors Applications Symposium (SAS)*; IEEE: New York, NY, USA, 2018; pp. 1–6.
149. Dong, Y.; Min, X.; Kim, W.S. A 3-D-Printed Integrated PCB-Based Electrochemical Sensor System. *IEEE Sens. J.* **2018**, *18*, 2959–2966. [CrossRef]
150. Katseli, V.; Economou, A.; Kokkinos, C. A novel all-3D-printed cell-on-a-chip device as a useful electroanalytical tool: Application to the simultaneous voltammetric determination of caffeine and paracetamol. *Talanta* **2020**, *208*, 120388. [CrossRef]
151. Zhang, X.; Wasserberg, D.; Breukers, C.; Connell, B.J.; Schipper, P.J.; van Dalum, J.; Baeten, E.; van den Blink, D.; Bloem, A.C.; Nijhuis, M.; et al. An inkjet-printed polysaccharide matrix for on-chip sample preparation in point-of-care cell counting chambers. *RSC Adv.* **2020**, *10*, 18062–18072. [CrossRef]

152. Syedmoradi, L.; Daneshpour, M.; Alvandipour, M.; Gomez, F.a.; Hajghassem, H.; Omidfar, K. Point of care testing: The impact of nanotechnology. *Biosens. Bioelectron.* **2017**, *87*, 373–387. [CrossRef]
153. Uliana, C.V.; Peverari, C.R.; Afonso, A.S.; Cominetti, M.R.; Faria, R.C. Fully disposable microfluidic electrochemical device for detection of estrogen receptor alpha breast cancer biomarker. *Biosens. Bioelectron.* **2018**, *99*, 156–162. [CrossRef] [PubMed]
154. Cinti, S.; Minotti, C.; Moscone, D.; Palleschi, G.; Arduini, F. Fully integrated ready-to-use paper-based electrochemical biosensor to detect nerve agents. *Biosens. Bioelectron.* **2017**, *93*, 46–51. [CrossRef]
155. Perumal, V.; Hashim, U. Advances in biosensors: Principle, architecture and applications. *J. Appl. Biomed.* **2014**, *12*, 1–15. [CrossRef]
156. de León, S.E.; Pupovac, A.; McArthur, S.L. Three-Dimensional (3D) cell culture monitoring: Opportunities and challenges for impedance spectroscopy. *Biotechnol. Bioeng.* **2020**, *17*, 1230–1240. [CrossRef] [PubMed]
157. Vlăsceanu, G.M.; Iovu, H.; Ioniţă, M. Graphene inks for the 3D printing of cell culture scaffolds and related molecular arrays. *Compos. Part B Eng.* **2019**, *162*, 712–723. [CrossRef]
158. Piro, B.; Mattana, G.; Reisberg, S. Transistors for chemical monitoring of living cells. *Biosensors* **2018**, *8*, 65. [CrossRef] [PubMed]
159. Mujica, M.L.; Gallay, P.A.; Perrachione, F.; Montemerlo, A.E.; Tamborelli, L.A.; Vaschetti, V.; Reartes, D.; Bollo, S.; Rodriguez, M.C.; Dalmasso, P.D.; et al. New trends in the development of electrochemical biosensors for the quantification of microRNAs. *J. Pharm. Biomed. Anal.* **2020**, *189*, 113478. [CrossRef] [PubMed]
160. Li, F.; Zhou, Y.; Yin, H.; Ai, S. Recent advances on signal amplification strategies in photoelectrochemical sensing of microRNAs. *Biosens. Bioelectron.* **2020**, *166*, 112476. [CrossRef] [PubMed]
161. Santhanam, M.; Algov, I.; Alfonta, L. DNA/RNA electrochemical biosensing devices a future replacement of PCR methods for a fast epidemic containment. *Sensors* **2020**, *20*, 4648. [CrossRef]
162. Li, F.; Li, Q.; Zuo, X.; Fan, C. DNA framework-engineered electrochemical biosensors. *Sci. China Life Sci.* **2020**, *63*, 1130–1141. [CrossRef] [PubMed]
163. Trotter, M.; Borst, N.; Thewes, R.; von Stetten, F. Review: Electrochemical DNA sensing—Principles, commercial systems, and applications. *Biosens. Bioelectron.* **2020**, *154*, 112069. [CrossRef]
164. Vasilescu, A.; Nunes, G.; Hayat, A.; Latif, U.; Marty, J.-L. Electrochemical affinity biosensors based on disposable screen-printed electrodes for detection of food allergens. *Sensors* **2016**, *16*, 1863. [CrossRef]
165. Iniesta, J.; García-Cruz, L.; Gomis-Berenguer, A.; Ania, C.O. Carbon materials based on screen-printing electrochemical platforms in biosensing applications. *SPR Electrochem.* **2016**, *13*, 133–169.
166. Xu, M.; Yadavalli, V.K. Flexible Biosensors for the Impedimetric Detection of Protein Targets Using Silk-Conductive Polymer Biocomposites. *ACS Sensors* **2019**, *4*, 1040–1047. [CrossRef]
167. Rathee, K.; Dhull, V.; Dhull, R.; Singh, S. Biosensors based on electrochemical lactate detection: A comprehensive review. *Biochem. Biophys. Reports* **2016**, *5*, 35–54. [CrossRef]
168. Mohanraj, J.; Durgalakshmi, D.; Rakkesh, R.A. Review-Current Trends in Disposable Graphene-Based Printed Electrode for Electrochemical Biosensors. *J. Electrochem. Soc.* **2020**, *167*, 067523.
169. Sonawane, A.; Manickam, P.; Bhansali, S. Stability of Enzymatic Biosensors for Wearable Applications. *IEEE Rev. Biomed. Eng.* **2017**, *10*, 174–186.
170. Sin, M.L.Y.; Mach, K.E.; Wong, P.K.; Liao, J.C. Advances and challenges in biosensor-based diagnosis of infectious diseases. *Expert Rev. Mol. Diagn.* **2014**, *14*, 225–244. [CrossRef]
171. Tonello, S.; Giorgi, G.; Pisu, S.; Cester, A. Organic substrates for novel printed sensors in neural interfacing: A measurement method for cytocompatibility analysis. In *2020 IEEE International Symposium on Medical Measurements and Applications (MeMeA)*; IEEE: New York, NY, USA, 2020; pp. 1–6.
172. Courtney, J.; Woods, E.; Scholz, D.; Hall, W.W.; Gautier, V.W. MATtrack: A MATLAB-Based Quantitative Image Analysis Platform for Investigating Real-Time Photo-Converted Fluorescent Signals in Live Cells. *PLoS ONE* **2015**, *10*, e0140209. [CrossRef]
173. Czech, E.; Aksoy, B.A.; Aksoy, P.; Hammerbacher, J. Cytokit: A single-cell analysis toolkit for high dimensional fluorescent microscopy imaging. *BMC Bioinform.* **2019**, *20*, 448. [CrossRef]
174. Chawla, K.; Burgel, S.C.; Schmidt, G.W.; Kaltenbach, H.; Rudolf, F.; Frey, O.; Hierlemann, A. Integrating impedance-based growth-rate monitoring into a microfluidic cell culture platform for live-cell microscopy. *Microsyst. Nanoeng.* **2018**, *4*, 8. [CrossRef]

175. Mayer, C.R.; Arsenovic, P.T.; Bathula, K.; Denis, K.B.; Conway, D.E. Characterization of 3D Printed Stretching Devices for Imaging Force Transmission in Live-Cells. *Cell. Mol. Bioeng.* **2019**, *12*, 289–300. [CrossRef] [PubMed]
176. Khan, M.Z.H.; Hasan, M.R.; Hossain, S.I.; Ahommed, M.S.; Daizy, M. Ultrasensitive detection of pathogenic viruses with electrochemical biosensor: State of the art. *Biosens. Bioelectron.* **2020**, *166*, 112431. [CrossRef]
177. Cesewski, E.; Johnson, B.N. Electrochemical biosensors for pathogen detection. *Biosens. Bioelectron.* **2020**, *159*, 112214. [CrossRef] [PubMed]
178. Alafeef, M.; Moitra, P.; Pan, D. Nano-enabled sensing approaches for pathogenic bacterial detection. *Biosens. Bioelectron.* **2020**, *165*, 112276. [CrossRef]
179. Poma, N.; Vivaldi, F.M.; Bonini, A.; Salvo, P.; Melai, B.; Bottai, D.; Tavanti, A.; di Francesco, F. A graphenic and potentiometric sensor for monitoring the growth of bacterial biofilms. *Sens. Actuators B Chem.* **2020**, *323*. [CrossRef]
180. Bao, C.; Kim, W.S. Perspective of Printed Solid-State Ion Sensors toward High Sensitivity and Selectivity. *Adv. Eng. Mater.* **2020**, *22*. [CrossRef]
181. Moulahoum, H.; Ghorbanizamani, F.; Zihnioglu, F.; Turhan, K.; Timur, S. How should diagnostic kits development adapt quickly in COVID 19-like pandemic models? Pros and cons of sensory platforms used in COVID-19 sensing. *Talanta* **2020**, *222*, 121534. [CrossRef]
182. Afzal, A. Molecular diagnostic technologies for COVID-19: Limitations and challenges. *J. Adv. Res.* **2010**.
183. Rasheed, P.A.; Sandhyarani, N. Electrochemical DNA sensors based on the use of gold nanoparticles: A review on recent developments. *Microchim. Acta* **2017**, *184*, 981–1000. [CrossRef]
184. Kaushik, M.; Khurana, S.; Mehra, K.; Yadav, N.; Mishra, S.; Kukreti, S. Emerging trends in advanced nanomaterials based electrochemical genosensors. *Curr. Pharm. Des.* **2018**, *24*, 3697–3709. [CrossRef] [PubMed]
185. Moccia, M.; Caratelli, V.; Cinti, S.; Pede, B.; Avitabile, C.; Saviano, M.; Imbriani, A.L.; Moscone, D.; Arduini, F. Paper-based electrochemical peptide nucleic acid (PNA) biosensor for detection of miRNA-492: A pancreatic ductal adenocarcinoma biomarker. *Biosens. Bioelectron.* **2020**, *165*, 112371. [CrossRef]
186. Erdem, A.; Eksin, E. ZNA probe immobilized single-use electrodes for impedimetric detection of nucleic acid hybridization related to single nucleotide mutation. *Anal. Chim. Acta* **2019**, *1071*, 78–85. [CrossRef]
187. Grieshaber, D.; MacKenzie, R.; Vörös, J.; Reimhult, E. Electrochemical Biosensors—Sensor Principles and Architectures. *Sensors* **2008**, *8*, 1400–1458. [CrossRef]
188. Jimenez-Falcao, S.; Parra-Nieto, J.; Pérez-Cuadrado, H.; Martínez-Máñez, R.; Martínez-Ruiz, P.; Villalonga, R. Avidin-gated mesoporous silica nanoparticles for signal amplification in electrochemical biosensor. *Electrochem. Commun.* **2019**, *108*, 106556. [CrossRef]
189. Arrabito, G.; Ferrara, V.; Ottaviani, A.; Cavaleri, F.; Cubisino, S.; Cancemi, P.; Ho, Y.; Knudsen, B.R.; Hede, M.S.; Pellerito, C.; et al. Imbibition of Femtoliter-Scale DNA-Rich Aqueous Droplets into Porous Nylon Substrates by Molecular Printing. *Langmuir* **2019**, *35*, 17156–17165. [CrossRef]
190. Pantazis, A.K.; Papadakis, G.; Parasyris, K.; Stavrinidis, A.; Gizeli, E. 3D-printed bioreactors for DNA amplification: Application to companion diagnostics. *Sens. Actuators B Chem.* **2020**, *319*, 128161. [CrossRef]
191. Song, Y.; Gyarmati, P. Rapid DNA detection using filter paper. *N. Biotechnol.* **2020**, *55*, 77–83. [CrossRef]
192. Nguyen, H.V.; Nguyen, V.D.; Lee, E.Y.; Seo, T.S. Point-of-care genetic analysis for multiplex pathogenic bacteria on a fully integrated centrifugal microdevice with a large-volume sample. *Biosens. Bioelectron.* **2019**, *136*, 132–139. [CrossRef]
193. Martínez-Domingo, C.; Conti, S.; de la Escosura-Muñiz, A.; Terés, L.; Merkoçi, A.; Ramon, E. Organic-based field effect transistors for protein detection fabricated by inkjet-printing. *Org. Electron.* **2020**, *84*, 105794. [CrossRef]
194. Jaeger, J.; Groher, F.; Stamm, J.; Spiehl, D.; Braun, J.; Dorsam, E.; Suess, B. Characterization and inkjet printing of an RNA aptamer for paper-based biosensing of ciprofloxacin. *Biosensors* **2019**, *9*, 7. [CrossRef]
195. Nag, A.; Afsrimanesh, N.; Mukhopadhyay, S.C. Impedimetric microsensors for biomedical applications. *Curr. Opin. Biomed. Eng.* **2019**, *9*, 1–7. [CrossRef]
196. Ameri, M.; Shabaninejad, Z.; Movahedpour, A.; Sahebkar, A.; Mohammadi, S.; Hosseindoost, S.; Ebrahimi, M.S.; Savardashtaki, A.; Karimipour, M.; Mirzaei, H. Biosensors for detection of Tau protein as an Alzheimer's disease marker. *Int. J. Biol. Macromol.* **2020**, *162*, 1100–1108. [CrossRef] [PubMed]

197. Sharafeldin, M.; Kadimisetty, K.; Bhalerao, K.S.; Chen, T.; Rusling, J.F. 3D-printed immunosensor arrays for cancer diagnostics. *Sensors* **2020**, *20*, 4514. [CrossRef]
198. Kashefi-Kheyrabadi, L.; Kim, J.; Chakravarty, S.; Park, S.; Gwak, H.; Kim, S.; Mohammadniaei, M.; Lee, M.; Hyun, K.; Jung, H. Detachable microfluidic device implemented with electrochemical aptasensor (DeMEA) for sequential analysis of cancerous exosomes. *Biosens. Bioelectron.* **2020**, *169*, 112622. [CrossRef] [PubMed]
199. Li, Z.; Xue, Q.; Wang, Q.; Zhang, H.; Duan, X. Biomolecules Detection Using Microstrip Sensor with Highly-ordered Nanowires Array. In *Proceedings of IEEE Sensors*; IEEE: New York, NY, USA, 2019; Volume 2019.
200. Martín, C.M.; Pedrero, M.; Gamella, M.; Montero-Calle, A.; Barderas, R.; Campuzano, S.; Pingarron, J.M. A novel peptide-based electrochemical biosensor for the determination of a metastasis-linked protease in pancreatic cancer cells. *Anal. Bioanal. Chem.* **2020**, 1–12. [CrossRef]
201. Sardesai, A.U.; Dhamu, V.N.; Paul, A.; Muthukumar, S.; Prasad, S. Design and electrochemical characterization of spiral electrochemical notification coupled electrode (SENCE) platform for biosensing application. *Micromachines* **2020**, *11*, 333. [CrossRef]
202. Damiati, S.; Haslam, C.; Sopstad, S.; Peacock, M.; Whitley, T.; Davey, P.; Awan, S.A. Sensitivity Comparison of Macro- and Micro-Electrochemical Biosensors for Human Chorionic Gonadotropin (hCG) Biomarker Detection. *IEEE Access* **2019**, *7*, 94048–94058. [CrossRef]
203. Kanitthamniyom, P.; Zhou, A.; Feng, S.; Liu, A.; Vasoo, S.; Zhang, Y. A 3D-printed modular magnetic digital microfluidic architecture for on-demand bioanalysis. *Microsyst. Nanoeng.* **2020**, *6*, 1–11. [CrossRef]
204. Kit-Anan, W.; Olarnwanich, A.; Sriprachuabwong, C.; Karuwan, C.; Tuantranont, A.; Wisitsoraat, A.; Srituravanich, W.; Pimpin, A. Disposable paper-based electrochemical sensor utilizing inkjet-printed Polyaniline modified screen-printed carbon electrode for Ascorbic acid detection. *J. Electroanal. Chem.* **2012**, *685*, 72–78. [CrossRef]
205. Arshad, R.; Rhouati, A.; Hayat, A.; Nawaz, M.H.; Yameen, M.A.; Mujahid, A.; Latif, U. MIP-Based Impedimetric Sensor for Detecting Dengue Fever Biomarker. *Appl. Biochem. Biotechnol.* **2020**, *191*, 1384–1394. [CrossRef] [PubMed]
206. Dunajová, A.A.; Gal, M.; Tomvcikova, K.; Sokolova, R.; Kolivovska, V.; Vanvevckova, E.; Kielar, F.; Kostolansky, F.; Varevckova, E.; Naumowicz, M. Ultrasensitive impedimetric imunosensor for influenza A detection. *J. Electroanal. Chem.* **2020**, *858*, 113813. [CrossRef]
207. Lien, T.T.N.; Takamura, Y.; Tamiya, E.; Vestergaard, M.C. Modified screen printed electrode for development of a highly sensitive label-free impedimetric immunosensor to detect amyloid beta peptides. *Anal. Chim. Acta* **2015**, *892*, 69–76. [CrossRef]
208. Degefa, T.H.; Hwang, S.; Kwon, D.; Park, J.H.; Kwak, J. Aptamer-based electrochemical detection of protein using enzymatic silver deposition. *Electrochim. Acta* **2009**, *54*, 6788–6791. [CrossRef]
209. Patel, S.; Nanda, R.; Sahoo, S.; Mohapatra, E. Biosensors in Health Care: The Milestones Achieved in Their Development towards Lab-on-Chip-Analysis. *Biochem. Res. Int.* **2016**, *2016*, 3130469. [CrossRef] [PubMed]
210. Maier, D.; Laubender, E.; Basavanna, A.; Schumann, S.; Güder, F.; Urban, G.A.; Dincer, C. Toward Continuous Monitoring of Breath Biochemistry: A Paper-Based Wearable Sensor for Real-Time Hydrogen Peroxide Measurement in Simulated Breath. *ACS Sens.* **2019**, *4*, 2945–2951. [CrossRef] [PubMed]
211. Hondred, J.A.; Breger, J.C.; Alves, N.J.; Trammell, S.A.; Walper, S.A.; Medintz, I.L.; Claussen, J.C. Printed Graphene Electrochemical Biosensors Fabricated by Inkjet Maskless Lithography for Rapid and Sensitive Detection of Organophosphates. *ACS Appl. Mater. Interfaces* **2018**, *10*, 11125–11134. [CrossRef] [PubMed]
212. Cao, Q.; Liang, B.; Yu, C.; Fang, L.; Tu, T.; Wei, J.; Ye, X. High accuracy determination of multi metabolite by an origami-based coulometric electrochemical biosensor. *J. Electroanal. Chem.* **2020**, *873*, 114358. [CrossRef]
213. Murastov, G.; Bogatova, E.; Brazovskiy, K.; Amin, I.; Lipovka, A.; Dogadina, E.; Cherepnyov, A.; Ananyeva, A.; Plotnikov, E.; Ryabov, V.; et al. Flexible and water-stable graphene-based electrodes for long-term use in bioelectronics. *Biosens. Bioelectron.* **2020**, *166*, 112426. [CrossRef]
214. Mejía-Salazar, J.R.; Cruz, K.R.; Vásques, E.M.M.; Jr, O.N.d. Microfluidic point-of-care devices: New trends and future prospects for ehealth diagnostics. *Sensors* **2020**, *20*, 1951. [CrossRef]
215. Palenzuela, C.L.M.; Pumera, M. (Bio)Analytical chemistry enabled by 3D printing: Sensors and biosensors. *TrAC—Trends Anal. Chem.* **2018**, *103*, 110–118. [CrossRef]
216. Lamas-Ardisana, P.J.; Martínez-Paredes, G.; Añorga, L.; Grande, H.J. Glucose biosensor based on disposable electrochemical paper-based transducers fully fabricated by screen-printing. *Biosens. Bioelectron.* **2018**, *109*, 8–12. [CrossRef]

217. Dayakar, T.; Rao, K.V.; Bikshalu, K.; Malapati, V.; Sadasivuni, K.K. Non-enzymatic sensing of glucose using screen-printed electrode modified with novel synthesized CeO$_2$@CuO core shell nanostructure. *Biosens. Bioelectron.* **2018**, *111*, 166–173. [CrossRef]
218. Espro, C.; Marini, S.; Giusi, D.; Ampelli, C.; Neri, G. Non-enzymatic screen printed sensor based on Cu2O nanocubes for glucose determination in bio-fermentation processes. *J. Electroanal. Chem.* **2020**, *873*, 114354. [CrossRef]
219. Criscuolo, F.; Cantu, F.; Taurino, I.; Carrara, S.; Micheli, G.D. Flexible sweat sensors for non-invasive optimization of lithium dose in psychiatric disorders. In *Proceedings of IEEE Sensors*; IEEE: New York, NY, USA, 2019; Volume 2019.
220. Demuru, S.; Kunnel, B.P.; Briand, D. Real-Time Multi-Ion Detection in the Sweat Concentration Range Enabled by Flexible, Printed, and Microfluidics-Integrated Organic Transistor Arrays. *Adv. Mater. Technol.* **2020**. [CrossRef]
221. Kang, T.-H.; Lee, S.; Hwang, K.; Shim, W.; Lee, K.; Lim, J.; Yu, W.; Choi, I.; Yi, H. All-Inkjet-Printed Flexible Nanobio-Devices with Efficient Electrochemical Coupling Using Amphiphilic Biomaterials. *ACS Appl. Mater. Interfaces* **2020**, *12*, 24231–24241. [CrossRef]
222. Cao, Q.; Liang, B.; Tu, T.; Wei, J.; Fang, L.; Ye, X. Three-dimensional paper-based microfluidic electrochemical integrated devices (3D-PMED) for wearable electrochemical glucose detection. *RSC Adv.* **2019**, *9*, 5674–5681. [CrossRef]
223. Noviana, E.; McCord, C.P.; Clark, K.M.; Jang, I.; Henry, C.S. Electrochemical paper-based devices: Sensing approaches and progress toward practical applications. *Lab Chip* **2020**, *20*, 9–34. [CrossRef]
224. Munaz, A.; Vadivelu, R.K.; John, J.S.; Barton, M.; Kamble, H.; Nguyen, N.-T. Three-dimensional printing of biological matters. *J. Sci. Adv. Mater. Devices* **2016**, *1*, 1–17. [CrossRef]
225. Schmatz, B.; Lang, A.W.; Reynolds, J.R. Fully Printed Organic Electrochemical Transistors from Green Solvents. *Adv. Funct. Mater.* **2019**, *29*, 1905266. [CrossRef]
226. Capella, J.V.; Bonastre, A.; Campelo, J.C.; Ors, R.; Peris, M. IoT & environmental analytical chemistry: Towards a profitable symbiosis. *Trends Environ. Anal. Chem.* **2020**, *27*, e00095.

Publisher's Note: MDPI stays neutral with regard to jurisdictional claims in published maps and institutional affiliations.

© 2020 by the authors. Licensee MDPI, Basel, Switzerland. This article is an open access article distributed under the terms and conditions of the Creative Commons Attribution (CC BY) license (http://creativecommons.org/licenses/by/4.0/).

Review

Printed Circuit Board (PCB) Technology for Electrochemical Sensors and Sensing Platforms

Hamed Shamkhalichenar [1], Collin J. Bueche [2] and Jin-Woo Choi [2,3,*]

1. Aquatic Germplasm and Genetic Resources Center, School of Renewable Natural Resources, Louisiana State University Agricultural Center, Baton Rouge, LA 70820, USA; hshamk1@lsu.edu
2. School of Electrical Engineering and Computer Science, Louisiana State University, Baton Rouge, LA 70803, USA; cbuech8@lsu.edu
3. Center for Advanced Microstructures and Devices, Louisiana State University, Baton Rouge, LA 70803, USA
* Correspondence: choijw@lsu.edu

Received: 18 September 2020; Accepted: 27 October 2020; Published: 30 October 2020

Abstract: The development of various biosensors has revolutionized the healthcare industry by providing rapid and reliable detection capability. Printed circuit board (PCB) technology has a well-established industry widely available around the world. In addition to electronics, this technology has been utilized to fabricate electrical parts, including electrodes for different biological and chemical sensors. High reproducibility achieved through long-lasting standard processes and low-cost resulting from an abundance of competitive manufacturing services makes this fabrication method a prime candidate for patterning electrodes and electrical parts of biosensors. The adoption of this approach in the fabrication of sensing platforms facilitates the integration of electronics and microfluidics with biosensors. In this review paper, the underlying principles and advances of printed board circuit technology are discussed. In addition, an overview of recent advancements in the development of PCB-based biosensors is provided. Finally, the challenges and outlook of PCB-based sensors are elaborated.

Keywords: printed circuit board; sensor electrode; electrochemical sensor

1. Introduction

In general, the primary factors in choosing a desirable detection method and tools for specific applications are cost, sensitivity, reliability, and rapidity. Cost is one of the main driving forces behind modern innovation but not the only important parameter. Obtaining reliable and accurate measurements in a short amount of time cannot always be sacrificed to reduce expenses. In clinical diagnostic applications, the reliability and rapidity of data play important roles. For example, reliable real-time measurement of blood glucose is essential in controlling the progress of diabetes [1].

To develop cost-effective and accurate sensors, the adoption of suitable detection methods, fabrication techniques, and materials required for the development of the sensors should be considered. Electrochemical analyses can offer an economical approach to quantify chemicals and detect changes in the physical characteristic of materials with high selectivity and sensitivity [2]. In terms of equipment, such techniques generally require electrochemical sensors composed of two or three electrodes called working (sensing), reference, and counter (auxiliary) electrodes in addition to electronic instrumentation for collecting data. Although the conventionally required electronic instrumentation can be sizeable and expensive, these devices can be miniaturized using recent advances in electronics. The implementation of such miniaturized instruments can facilitate the utilization of electrochemical sensors in point-of-care and field-deployable applications.

Different fabrication methods can be considered to construct electrochemical sensors. For example, microfabrication techniques used in the semiconductor industry are well established due to their flexibility in the adoption of a vast range of materials and techniques offering outstanding control over the sensors parameters. However, multiple techniques (e.g., sputtering, chemical vapor deposition, photolithography) and specialized facilities may be required to fabricate these sensors [3]. As an alternative approach, printed circuit board technology (PCB) has the potential for the construction of sensors. This technology is a well-established economical manufacturing method widely used to fabricate electronic circuitry. Nowadays, PCB fabrication is broadly available at relatively low cost due to considerable growth in the electronics industry during the past decades [4]. PCB technologies make it possible to pattern conductive electrodes with high precision that can be used as a substrate for sensors. Although PCB technology employs techniques similar to microfabrication processes, it provides widely available affordable manufacturing possibilities.

This review aims to provide an overview of the utilization of PCB technology in the development of electrochemical sensors and miniaturized sensing platforms. To begin with, the background of the printed circuit board technology is discussed, along with recent advances in this field. Next, an overview of the recent advances in the development of PCB-based electrochemical sensors and sensing platforms was provided.

2. Printed Circuit Board Technology

2.1. History

In 1903, Albert Hanson created the first printed circuit board by laminating flat foil conductors to an insulating board. In 1904, Thomas Edison formed conductors onto linen paper by utilizing patterned polymer adhesives. While the aforementioned designs would be nearly unrecognizable today, Hanson and Edison laid the groundwork for what would become an essential component to the modern electronics industry. From the early 1900s until the 1940s, few advancements were made, and the boards were limited to the usage of only a single side. However, the United States Army began to use PCBs to make proximity fuses in 1943, later releasing the technology to the public after the war [5]. In the 1950s, through-hole technology was the most popular method of mounting electronic components onto a board. Through-hole technology involves mounting the leads of the component in holes drilled on the board and soldering the leads in place from the underside of the board. Because of the need to drill holes into the board, the available space and routing area was always limited when manufacturing using through-hole technology.

Surface-mount technology (SMT) became a mainstream manufacturing technology for electronics on printed circuit boards in the 1980s, leading to a significant reduction in size, cost, and complexity. SMT allows for more components to be placed in the same space compared to through-hole technology due to the fact that no holes have to be drilled. Furthermore, components can be placed on both sides of the board. Most importantly, SMT boards can be fabricated in multiple layers, which makes them a great candidate for the implementation of high-speed electronics by providing precisive control over the impedance of the traces and electromagnetic interference. Regarding current PCB manufacturing technology, SMT is heavily favored over through-hole. However, through-hole is still used for simpler boards and is easier to solder by hand due to the larger size as opposed to SMT boards. The vast majority of boards manufactured today employ surface-mount technology, and anyone can utilize CAD software to create a design, send it to a manufacturer, and have their own PCB constructed.

2.2. Materials

The most common printed circuit board substrate is known as FR4 (flame retardant 4). FR4 is a class of materials that meet National Electrical Manufacturers Association Industrial Thermosetting Products (NEMA LI 1-1998) requirements. The basis of FR4 is composed of woven fiberglass cloth combined with an epoxy resin binder that is flame-resistant. FR4 has near-zero water absorption, as well as an excellent strength to weight ratio, and is an excellent insulator regardless of moisture levels in the ambient. Typically, the flame-resistant material in FR4 is bromine. Along with its other aforementioned properties, the reason FR4 is the most popular substrate because it is easy to manufacture and usually the cheapest material available. However, other materials are often used depending on the environment the board will be placed in, the budget, and the required circuit properties.

Polyimide laminates offer improvements in every category over FR4, most importantly, higher temperature performance, electrical performance, survivability, and resistance to expansion. The cost to manufacture polyimides, however, is higher than FR4. Teflon laminates offer improved electrical properties over both FR4 and polyimide-based substrates; however, they cost significantly more to produce than both and require specialized equipment and a highly-skilled workforce. Teflon laminates can be coated onto glass fabric or manufactured as an unsupported mesh, giving it an adaptability factor that neither the FR4 nor the polyimide possesses.

The multi-layer manufacturing process begins with the creation of a computer-aided design or CAD, which is then sent to the chosen manufacturer. The manufacturer checks to make sure the CAD is compatible with their equipment. A photographic image of the CAD is printed on film, and the image is transferred to the board surface, using photosensitive dry-film and ultraviolet light in a cleanroom. The photographic film is removed, and excess copper is etched from the board. The inner layers receive an oxide layer application and are then stacked with prepreg providing insulation between layers, and copper foil is added to the top and bottom of the newly created stack. An oxide layer application strengthens the laminate bond by increasing the roughness of clad copper. The oxide layer is a chemical composition consisting of compounds such as sodium chlorite ($NaClO_2$), water, and sodium hydroxide. The internal layers are laminated by subjecting them to extreme pressure and high temperature. Slowly, pressure is released, and the PCB cures while still at a high temperature. Next, holes are drilled to secure the stack, and excess copper is filed off. A chemical is used to fuse all the layers of the board together, and then the board is cleaned. After cleaning, a series of chemicals bathe the board, resulting in a layer of copper weighing 1 oz/ft^2 (305.152 g/m^2), which results in a thickness of 1.4 mil (35 μm), filling in the drill holes and settling on the top layer. Using imperial units such as oz, mil, oz/ft^2 over metric units is a convention in PCB industry. In addition, oz is often used over oz/ft^2 to refer copper weight spread evenly over 1 ft^2 (305.152 g/m^2) PCB area to determine the copper layer thickness. Once again, the board needs to receive a photoresist application, but only on the outer layers. After the photoresist application, the outer layers are plated the exact same way the inner layers were, but a plating of tin is applied to protect the outer-layer copper from etching. Etching takes place on the outer layers, and excess copper is removed via a copper etchant, with the tin safeguarding the remaining copper. The panels are cleaned and prepared for a solder mask [6]. After cleaning, ink epoxy and solder mask film are applied, and the boards are exposed to ultraviolet light to designate a certain area of the solder mask for removal. The board is baked, allowing the solder mask to cure (Figure 1). The board is plated with gold, silver, or hot air solder level (HASL), enabling the components to be soldered to the pads and to protect the copper. The process by which the board receives plating is known as electroless nickel immersion gold (ENIG). A nickel layer is applied to the copper as a diffusion barrier. Following the nickel layer is a thin gold layer which serves to prevent nickel oxidation and maintains a solid surface of which to solder [7]. After gold or silver-plating, the board is silk-screened, receiving all of the vital information, such as warning labels and company ID numbers. Finally, the board is tested and cut to fit design specifications.

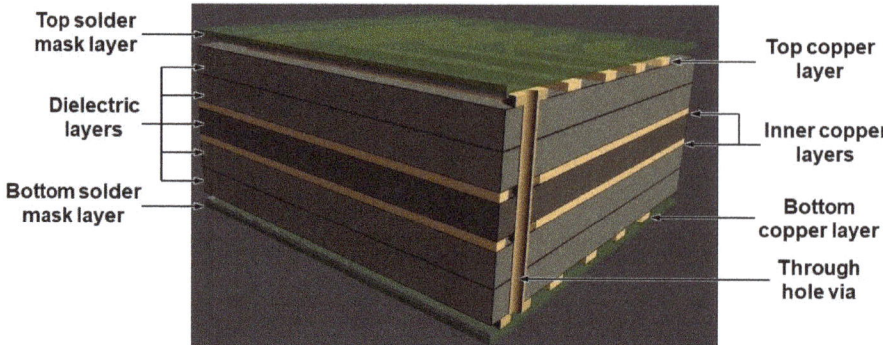

Figure 1. A 3-D schematic design of a multi-layer printed circuit board (PCB) board composed of four copper layers (two internal layers), five dielectric layers, and the top and bottom mask layer.

Finishes on the board surface protect exposed copper and provide solderable surfaces. Historically, HASL has been the most prevalent finish in the industry. HASL costs little and is widely available. The circuit boards are immersed in a molten mixture of tin/lead, and excess solder is removed by blowing hot air across the surface of the board. However, the use of ENIG has been rapidly increasing. Nickel forms the layer, which provides a barrier for the copper, as well as being the solderable surface. Gold is then used to protect the nickel and provides the low contact resistance necessary for the thin gold deposits. Electroless nickel electroless palladium immersion gold (ENEPIG), which has been developed relatively recently, has seen increasing usage, despite its high cost because of reliance on palladium [8]. ENEPIG is significantly more resistant to corrosion compared with ENIG and HASL, allowing the PCB to last for a longer period. Across all measurable categories, ENEPIG is superior to HASL and ENIG, but the cost is also noticeably higher. Furthermore, of note is the hard electrolytic gold finish, which consists of a layer of gold plated over a coat of nickel. Hard gold is very durable and used on boards that experience high wear. Hard electrolytic gold is similar to ENIG, but the hard gold layer is generally two to three times as thick (0.005–0.010 mil or 0.127–0.254 µm). In regard to high-wear areas of the board that use the hard gold as a protective layer, the gold can be as thick as 1 mil (25.4 µm), meaning hard electrolytic gold plating is an expensive option.

2.3. State-of-the-Art Technology

As PCB technology has progressed over the years, a few key features have gradually improved as well. The most important of these are the number of conducting layers, minimum trace spacing, and minimum trace width. The maximum number of layers most manufacturers will produce is 40. Trace width and spacing are directly proportional to the weight of copper that plates the board. With only 0.5 oz (14.17 g) of copper on the inner layers, trace width can be minimized to 4 mil (101.6 µm). A 1 oz (28.35 g) copper deposit on the outer layers will yield the same 4 mil (101.6 µm) trace width. Distance between traces is identical to width for the majority of manufacturers.

Many applications require cyclic movement or stretching while still maintaining an electrical connection. To fit this need, flexible circuit boards have become an ever-evolving solution. Flexible PCBs or fPCBs have a variety of real-world applications, ranging from laptops to smartphones, engine management units to hard drives. Flexible PCBs are placed in laptops to ensure the connection between the computer and monitor remains intact, as a laptop may be folded thousands of times in its lifetime. Hard drives need to withstand high temperatures and transfer data quickly. Flexible PCBs can also be used as sensors, as on an automobile or even for general purpose. The automobile has many moving parts, and cars today have sensors on every conceivable component. A typical circuit board would not be able to withstand the stretching and bending or the constant change in ambient conditions.

Flexible circuit boards first came about in the early 2000s, with polydimethylsiloxane or PDMS as the most common substrate [5]. However, the most popular substrate materials have since become polyimide film, polyester film, and polyethylene naphthalate (PEN). Polyimide film is the most popular because of its great thermal resistance capabilities and excellent mechanical and electrical properties. High humidity absorption and proneness to tearing harms the polyimide film, but some variants have improved upon these areas. Offering competition is PEN, which fills an intermediate slot in the market. With most qualities' inferior to that of polyimides, PEN also comes in cheaper. The performance of PEN is still more than adequate for most applications, and it is becoming more and more popular each year. The smallest boards can be as thin as 4 mil (101.6 μm). Trace width on a flexible PCB can be as low as 3 mil (76.2 μm) and spacing as low as 3 mil (76.2 μm). Flex circuits can have multiple layers, up to 10 layers from most manufacturers. The lowest weight of copper is 0.5 oz (14.17 g) but ranges up to 2 oz (56.7 g).

3. PCB-Based Electrodes for Electrochemical Analyses

Copper is the most commonly used material in the fabrication of traces and electric contacts on PCB boards. However, the easy and unavoidable oxidation problem of copper limits its application in developing electrochemical sensors [9]. To overcome this challenge, a thin layer of inert metals, such as gold (Au) or platinum (Pt), can be deposited on the surface of PCB pads. The cyclic voltammetry analysis was performed using bare Cu PCB electrode shows a non-characteristic voltammogram since Cu can easily oxidize. On the other hand, the adoption of the Au-plated PCB electrode results in a stable voltammogram with a wide potential window acceptable for electrochemical biosensing applications (Figure 2). Different techniques can be employed to deposit gold on PCB electrodes, including electroless, electrodeposition, and sputtering. The PCB manufacturing services offer different surface finishes as part of the standard fabrication process. The most popular type is the ENIG coating [10]. However, the primary purpose of these coatings is to improve the solderability and shelf life of PCB boards and can leave exposed copper at the edges of the pads [11]. On the other hand, electrodeposition of hard Au or Pt can result in fully coated electrodes with a higher surface roughness, which can increase the effective surface area of the electrode [12].

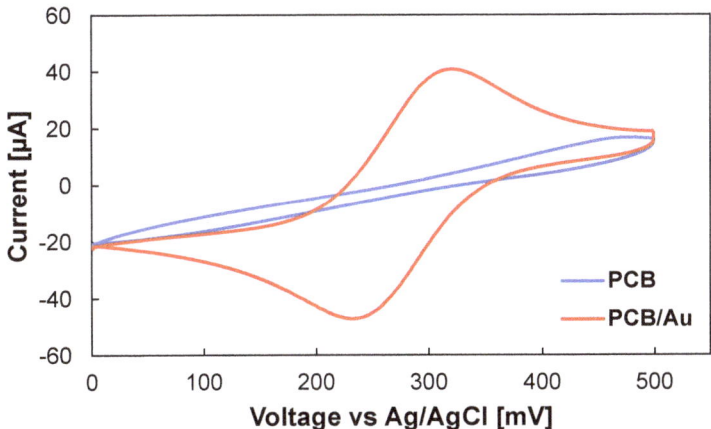

Figure 2. Cyclic voltammetry from −100 to 500 mV vs Ag/AgCl, inside 0.1 M KCl containing 5 mM $K_3Fe(CN)_6$ using PCB (blue line), and Au-plated PCB (red line) electrodes. The scan rate was 50 mV/s.

Specifically, for electrochemical sensors, the surface physical characteristics and chemical properties of sensing electrodes are of great importance in reliable and accurate detection of a target analyte. Additional Au electroplating can result in a pore-free surface and improved electron transfer at the surface of the electrode. However, based on an observation by Dutta et al. [13], after electroplating gold on the surface of PCB electrodes, exposed Cu and an organic layer with a high content of C and O may remain on the surface, which makes the electrode electrochemically inactive. They have suggested a cleaning process using acetone, ethanol, and water followed by ultrasonication in an aqueous solution containing ammonium hydroxide and hydrogen peroxide t make the electrodes electrochemically active and decrease the surface roughness originated from the organic layer (Figure 3).

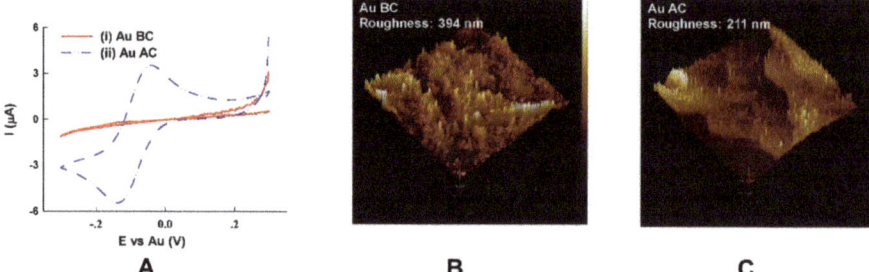

Figure 3. Comparison between Au-plated PCB electrode before and after cleaning: (**A**) cyclic voltammograms obtained from Au-Plated PCB electrodes in a PBS solution containing 4 mM $K_3Fe(CN)_6$ before cleaning (i) and after cleaning (ii); (**B**) atomic force microscopy of Au-plated PCB electrodes obtained before cleaning, and (**C**) after cleaning. The size of the atomic force microscopy (AFM) micrographs was not specified in the original figure. Reproduced from [13], Creative Commons Attribution License (http://creativecommons.org/licenses/by/4.0/).

Based on a study done by Evans et al. [14] on an Au-plated PCB electrode, the presence of chloride ions in the buffer solution could lead to the formation of a secondary electrochemical cell at the Au and copper interface. Comparing the electrodes' surface during the amperometry analysis using a buffer containing chloride ions (PBS) and without chloride compounds (HEPES) showed that the inclusion of chloride ions results in the reduction and corrosion of gold. A layer of electroplated nickel between gold and PCB copper pads can improve the adhesion of the gold [15]. Furthermore, the nickel layer acts as a diffusion barrier to reduce the penetration of copper through gold and avoids the copper reaching the surface and becoming oxidized [16]. Besides, a solder mask can be extended to cover electrode edges to avoid the exposure of the copper at the edges [12,17]. Using an interesting approach, a low temperature curing Au ink was screen printed to form an array of sensing electrodes (Figure 4) [18]. One of the important parameters in developing reliable sensing electrodes is the thickness of the plated gold, which has not been considered in many reported PCB-based biosensors. A thicker gold layer has been found to generate a more stable characteristic cyclic voltammogram [19].

In electrochemical biosensors, the reference electrode maintains its potential with minimum current passing through it. Silver chloride (Ag/AgCl) electrodes are one of the widely used reference electrodes in electrochemical analyses. A similar reference electrode can be integrated on PCB-based electrochemical biosensors by electroplating or electroless deposition, an additional Ag layer, followed by chlorination using HCl solution [18,20,21], sodium chloride solution [22], or sodium hypochlorite [23]. In some sensing applications in which the true reference potential is not necessary, a Pt or Au coated PCB pad can be used as a pseudo-reference electrode [24]. In addition, PCB electroless immersion silver plating is a standard industrial process offered by manufacturers, which can be adopted to be chlorinated and used as a reference electrode in biosensors [25]. To reduce the sensor size, a single electrode can act both as reference and counter; however, this causes higher noise levels in the measurements [22].

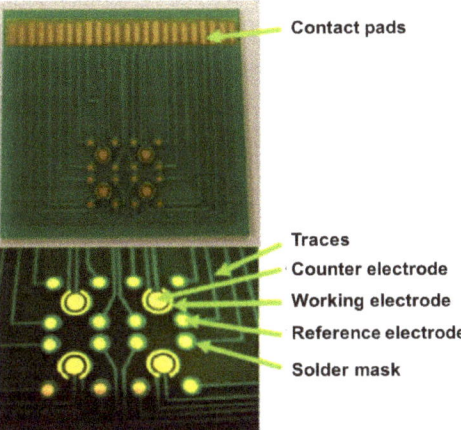

Figure 4. Photograph of the array chip fabricated using screen-printed Au ink on a PCB substrate. The diameter of the working and counter electrodes was 1 and 2 mm, respectively. The diameter of the outer and inner ring of the reference electrode was 4 and 3 mm. Reproduced with permission from [18], Copyright 2013 Elsevier.

4. Application of PCB-Based Electrochemical Sensors and Sensing Platforms

Various biosensors have been reported using PCB technology, which are summarized in Table 1. Glucose detection plays a key role in the diagnosis and management of diabetes mellitus. As a result, numerous enzymatic and non-enzymatic electrochemical sensors have been reported based on various fabrication methods, including screen printing [26–28], inkjet printing [29–31], and standard microfabrication processes [32–34]. Considering the availability and low manufacturing price of printed circuit boards, several PCB-based glucose biosensors have been reported recently. Glucose measurement is conventionally done using amperometry or cyclic voltammetry techniques through a three-electrode electrochemical cell.

Typically, the determination of glucose in a sample is done based on the glucose enzymatic reaction happening at the surface of the working electrode. The immobilization of glucose oxidase (GOx) enzyme on the surface of the sensing electrode affects the efficiency and sensitivity of the sensor. Although drop-casting the GOx on the Au-plated PCB electrode to develop glucose biosensor has been reported previously [15], more complex immobilization processes can improve the sensitivity and reproducibility of sensors. Dutta et al. [13] have formed a self-assembled monolayer (SAM) with activated carboxylic acid groups to covalently immobilize glucose oxidase on the surface of an Au-plated precleaned PCB electrode. A polymer matrix can also be used to immobilize the GOx on the sensing electrode. Kassanos et al. [20] used an additional layer of electropolymerized phenol red before drop-casting the GOx to develop an array of glucose-sensing PCB electrodes. After the immobilization process, the sensing electrode's surface was coated with a polyurethane film to improve the dynamic range of the sensor.

To improve the sensitivity and selectivity of the glucose biosensors, the sensing electrode surface can be modified by various nanomaterials [35]. Carbon-based nanomaterials have been widely adopted in electrochemical sensing application due to wide potential window, low background current, and improved electron transfer rate [36]. The dependency of carbon nanotubes' (CNTs) conductivity to surface absorbate and its ability to promote electron transfer have increased the use of these unique nanomaterials in developing a wide range of electrochemical biosensors [37]. Alhans et al. [38] drop-cast multi-walled and single-walled CNT dispersion solution on a PCB pad, which increased the electrochemical reactivity of the sensing electrode. The electrochemical impedance spectroscopy (EIS) results showed a decrease in electrodes resistance values and, consequently, a higher electron-transfer rate after the deposition of carbon nanotubes. The CNT working electrodes were modified by drop-casting GOx to form a low-cost PCB-based glucose biosensor.

Similarly, Li et al. [18] used a dispersion solution of CNT, polyvinylimidazole-Os (PVI-Os), enzyme (glucose or lactate oxidase), and chitosan composite sensing material to detect glucose and lactate electrochemically. Chitosan is a widely used biocompatible polymer in enzyme immobilization due to its high permeability toward the water and good adhesion [39]. As an electron mediator, PVI-Os improve the electron transfer while minimizing the enzyme leakage. The mentioned composite was dropped on an array of SAM-modified screen-printed Au electrodes to form glucose and lactate biosensors.

Table 1. The summary of the reported printed circuit board (PCB)-based electrochemical biosensors.

PCB Pads Modification	Sensing Electrode Surface Modification	Target Analyte	Detection Method	Ref.
Electroplated Au	GOx [1]	Glucose	Amperometry	[13]
Electroplated Ni, Au	GOx	Glucose	Cyclic voltammetry	[15]
Screen-printed Au	CNT [2], GOx/LOD	Glucose, Lactate	Amperometry	[18]
ENIG [3], Electroplated Au	Red phenol, GOx	Glucose	Amperometry	[20]
Au	CNT, GOx	Glucose	Amperometry and EIS [4]	[38]
Electroplated Ni, Au	Graphene, Au NPs [5], GOx	Glucose	Amperometry	[40]
Electroplated Au	Antibody	Mycobacterium tuberculosis	Amperometry	[41]
Electroplated Ni, Au	Antibody	Salmonella typhimurium	EIS	[22]
Electroplated Au	Antibody	Streptococcus mutans	EIS	[42]
Electroplated Au	-	Salmonella typhimurium	EIS	[43]
Electroplated Ni, Au	Antibody	Salmonella typhimurium	EIS	[44]
Electroplated Au	Antibody	IFN-γ [6]	Amperometry	[14]
Electroplated Au	Antibody	IFN-γ	Amperometry	[45]
Electroplated Ni, Au	Antibody	Interleukin-12	EIS	[23]
Electroplated Au	DNA probes	DNA	Sweep voltammetry	[46]
Electroplated Ni, Au	DNA probes	DNA	Square wave voltammetry	[47]
Electroplated Ni, Au	DNA probes	mRNA markers	Amperometry	[48]
Electroplated Au	ZnO, antibody	Troponin-T	EIS	[49]
Electroplated Ni, Au	-	Methylene blue	Cyclic voltammetry	[21]

[1] Glucose oxidase, [2] Carbon Nanotubes, [3] Electroless nickel immersion gold, [4] Electrochemical impedance spectroscopy, [5] Nanoparticles, [6] Interferon-gamma.

An important area that low-cost biological sensors have attracted attention is point-of-care (PoC) diagnostics of disease caused by various bacteria [50]. Tuberculosis (TB) is an infectious disease caused by Mycobacterium tuberculosis bacteria, which is considered a concerning global health-related issue [51]. Commercially fabricated PCB sensors can be employed to fabricate a low-cost biodetection system for PoC diagnosis of tuberculosis. Evans et al. [41] reported a PCB-based amperometric electrochemical sensor for the detection of tuberculosis using enzyme-linked immunosorbent assay (ELISA), which outperforms the standard colorimetric ELISA technique in terms of limit of detection. The working and counter electrodes were fabricated on the PCB, the reaction well was formed on top of the PCB using polymethyl methacrylate (PMMA), and an external Ag/AgCl reference electrode was introduced to the system (Figure 5). The capture antibodies were covalently localized on the Au-coated PCB sensing electrodes using thiol linkage. The detection of interferon-gamma (IFNγ) as a biomarker using the proposed ELISA system was done by electrochemical detection [41,52].

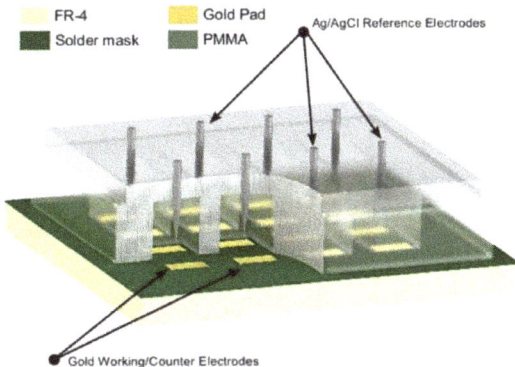

Figure 5. A three-dimensional representation of the fabricated prototype PCB-based sensor for the detection of tuberculosis using an enzyme-linked immunosorbent assay (ELISA). Reproduced from [41], Creative Commons Attribution License (http://creativecommons.org/licenses/by/4.0/).

According to the World Health Organization, foodborne illnesses caused by bacteria, such as *Salmonella typhimurium, Escherichia coli, Campylobacter,* and *Vibrio cholerae,* are a critical issue for public health [52]. Therefore, rapid and reliable detection of such pathogens plays a crucial role in the discovery of contaminations and controlling disease outbreaks. Nandakumar et al. [22] have developed a low cost PCB-based impedimetric sensor to detect *Salmonella typhimurium*. The sensing electrode was modified with *S. typhimurium*-specific antibodies. The infected samples can be distinguished by an increase in the impedance value resulted from the binding of the pathogens to the surface of the electrodes. A similar sensor structure was adopted by Dutta et al. [42] to detect *Streptococcus mutans* using a commercially fabricated PCB board.

Using a different approach, a PCB-based impedimetric sensor for the detection of *Salmonella* was reported by Wang et al. [43], in which the bacteria was selectively conjugated with magnetic and gold nanoparticles to form enzymatic bacteria. Next, the bacteria were employed to catalyze urea, which results in a decrease in the impedance of the sample.

One of the earliest applications of a PCB-based biological sensor was in molecular diagnostics, which was reported by the researchers at Motorola Inc. [46,53]. To electrochemically detect nucleic acids, the surface of the Au-plated PCB microarray was coated with DNA capture probes using a self-assembled monolayer (SAM). After the unlabeled nucleic acid targets were immobilized on this layer, ferrocene-modified nucleotides were introduced to the system as a signaling probe to form a sandwich complex. The SAM layer avoids non-specific binding of the electroactive species to the surface of the microarray while making the oxidation ferrocene-labeled adenosine derivative possible through the electron exchange with the gold. Later, this work led to a commercial sensor called eSensor®® produced by Motorola Inc. for nucleic acid target detection and genotyping. This sensor is composed of an array of gold working electrodes, a gold counter electrode, and an Ag/AgCl electrode, which can be accessed through connectors at the edge of the board. PCB-based DNA detection platforms with integrated microfluidic systems have been reported using this commercial PCB-based biological sensor [54,55].

Gassmann et al. [47] reported on a DNA detection chip with an integrated microfluidic system, which was capable of performing polymerase chain reaction (PCR). The PCR process was done by cycle heating using copper traces on the PCB based on the Joule heating concept. Two separate PCB boards, one with microchannels and the other with electrodes, were fabricated separately and stacked together to form the DNA chip. In addition to temperature cycling, Tseng et al. [21] incorporated a PCB-based sensor on their platform to detect methylene blue. During PCR amplification, the methylene blue concentration decreases due to the binding to double-stranded DNA (ds-DNA) and single-stranded DNA (ss-DNA). The concentration of the methylene blue can be monitored using cyclic voltammetry and the fabricated PCB-based electrochemical setup.

Another interesting application of a PCB-based sensor in molecular diagnostics was showcased by Acero Sánchez et al. [48] for breast cancer-related mRNA markers. The proposed platform was composed of a PCB array with 64 Au-coated individually addressable electrodes in conjunction with an integrated PMMA-based microfluidic system (Figure 6). The sensing electrode surface was cleaned using acetone, isopropanol, and water, followed by oxygen plasma treatment. The presence of O_2 removes the remaining organic materials, while the Au provides a fresh gold surface [56]. The integration of microfluid systems with PCB technology resulted in the emergence of the lab-on-PCB concept [57]. We suggest that interested readers refer to the review paper by Moschou et al. [58].

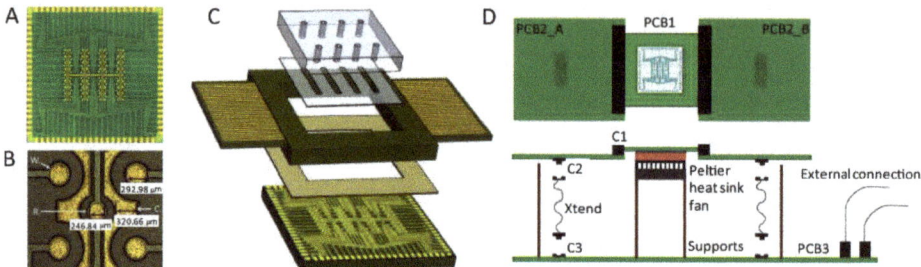

Figure 6. PCB based sensing platform developed for electrochemical detection of cancer-related mRNA markers: (**A**) Electrode arrayed developed using PCB technology; (**B**) magnified image of the PCB electrodes; (**C**) schematic the sensor along with the microfluidic system, and (**D**) fully assembled biosensing platform. The PCB-based chip was a square with a side length of 24.6 mm. The diameter of the working, reference, and counter electrodes was 300, 250, and 250 µm, respectively. Reproduced with permission from [48]. Copyright 2016 Elsevier.

Cytokine, interleukin-12 (IL-12), is a biomarker found to have elevated ranges in patients diagnosed with an autoimmune disease called multiple sclerosis (MS) [59]. Bhavsar et al. [23] developed a robust PCB-based sensor to detect this protein biomarker by immobilizing anti-IL-12 antibody on the surface of an Au-plated PCB sensing electrode and performing electrochemical impedance spectroscopy. The proposed method reduces the detection time to 90 s with an ultra-low limit of detection (<100 fM).

PCB sensing electrodes sputter-coated with zinc oxide (ZnO) has been reported to anchor capture antibodies [49]. Troponin-T is a cardiac biomarker that can be found in the bloodstream of patients with myocardial damage. The capture antibody was attached to the ZnO-modified PCB electrodes (Figure 7). The changes in the electrochemical impedance after capturing troponin-T by the capture antibody was used to detect the level of this protein.

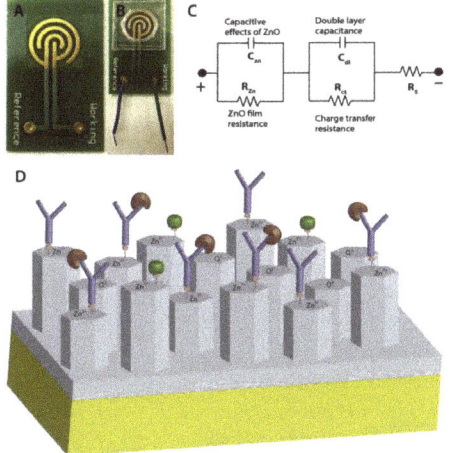

Figure 7. PCB based sensor developed for detection of Troponin-T as a cardiac biomarker: (**A**) PCB with electroplated gold electrodes and ZnO sputtered sensing site; (**B**) assembled sensor platform; polydimethylsiloxane (PDMS) manifold confines sample fluid on the ZnO sputtered sensing site; (**C**) electrical circuit model of the sensor, and (**D**) schematic of Troponin-T immunoassay on the nanocolumnar ZnO surface. The width of all the electrodes and the gap between each was 1 mm. Reproduced with permission from [49], Copyright 2016 Elsevier.

The detection of Troponin-I, another cardiac regulatory protein, using a PCB-based sensing platform, has been reported by Lee et al. [60]. However, the sensing unit itself was fabricated using standard microfabrication processes and attached to the PCB later. Although the fully PCB-based sensors provide advantages of integration of electronic measurement systems without implementing additional connection strategies, this work is one of the great examples to see how a sensing platform fabricated with different methods can be easily integrated into a PCB board. While the connection of the miniaturized sensor to the outside world remains challenging [61], PCB platforms can offer a reliable alternative approach to overcome this problem. Several studies have benefited from the advantages of PCB technology to provide an interface for their biological sensor to be connected to electronics [9,62–65].

Recently, flexible PCB technology has gained attention in the fabrication of biosensing platforms. The wide viability of manufacturing services, low weight, and mechanical flexibility of this technology make it a promising candidate to develop wearable biosensing devices. However, the flexible PCB can be adopted to develop miniaturized thin wearable measurement systems to be used in conjunction with microfabricated biosensors [66,67]. In addition, flexible PCB itself can be used as the backbone of the sensor. For example, Pu et al. [40] developed a glucose sensor for the detection of hypoglycemia in interstitial fluid (ISF) on a polyimide substrate using flexible PCB technology. They employed inkjet printing as an interesting approach to modify the electrode surface with graphene. The advantages that such technologies offer, in combination with a well-established and rapidly growing PCB manufacturing industry, make this approach a great alternative fabrication method for the implementation of various biosensors.

5. Conclusions and Outlook

PCB technology offers an alternative, low-cost approach for the fabrication of various sensors. This approach facilitates the transition of prototyped sensors to the market and end-users due to the preexisting manufacturing industry that advances rapidly. In addition, the integrability of fluidics and electronics with PCB-based biosensing platforms makes them a great candidate for standalone point-of-care diagnosis systems.

Although PCB technology shares a lot of fabrication methodologies with microfabrication processes, it offers additional advantages in terms of long-lasting and thriving PCB manufacturing industries. This facilitates the adoption of PCB-based biosensors by the market and end-users. Furthermore, the similarity in the process opens up new opportunities to adopt already investigated biosensors' designs and implement it on a PCB board to reduce the fabrication cost and promote commercialization possibilities. For example, novel materials, such as carbon nanotubes [68], reduced graphene oxide [69], metal nanomaterials [70], metal oxide nanoparticles [71], can be used to develop novel PCB-based biosensors with improved sensitivity and selectivity.

On the other hand, standard microfabrication processes outperform the PCB technology in terms of minimum feature size. However, the rapidly growing necessity for miniaturization of electronic systems pushes this industry toward improving this limitation. Besides, the use of copper, which is not usable in the electrochemical analysis, imposes additional modification steps for the development of reliable biosensors. To overcome this challenge, novel low-cost fabrication methods, such as inkjet printing [3] and screen printing [72], can be used along with PCB technology. In addition, further investigation into the electrochemical characterization of standard PCB pad finishes offered by the current industrial processes may lead to a promising substrate to perform electrochemical analysis.

Given the wide variety of target analytes and inherently different fabrication and detection methods utilized by the reviewed reports, the comparison between the sensors from a bioanalytical standpoint was beyond the scope of this review. However, interested readers can refer to multiple published review papers dedicated to electrochemical detection of a specific target analyte using specific detection methods and materials [50,73–76].

Overall, this review shows the capabilities of PCB technology as a reliable method to develop electrochemical sensors using different electroanalytical and bioanalytical approaches. The diminishing manufacturing price of PCBs due to the rapid growth of the electronic industry provides opportunities to adopt this technology for the fabrication of affordable disposable electrochemical sensors for point-of-care applications. Besides, the recent advancements in flexible PCB technology makes PCB-based sensors a promising candidate for detection in conditions that mechanical flexibility and total sensor weight is critical (e.g., wearable devices).

Author Contributions: Conceptualization, J.-W.C. and H.S.; methodology, J.-W.C. and H.S.; validation, J.-W.C., H.S. and C.J.B.; investigation, J.-W.C., H.S. and C.J.B.; data curation, J.-W.C., H.S. and C.J.B.; writing—original draft preparation, H.S. and C.J.B.; writing—review and editing, J.-W.C.; visualization, H.S.; supervision, J.-W.C.; project administration, J.-W.C. All authors have read and agreed to the published version of the manuscript.

Funding: This research was supported in part by funding from the National Institutes of Health, Office of Research Infrastructure Programs (R24-OD010441), and the President's Future Leaders in Research program of Louisiana State University.

Acknowledgments: We thank Terrence R. Tiersch for discussions. This manuscript was approved for publication by the Louisiana State University Agricultural Center as number 2020-241-34842.

Conflicts of Interest: The authors declare no conflict of interest.

References

1. Bantle, J.P.; Thomas, W. Glucose measurement in patients with diabetes mellitus with dermal interstitial fluid. *J. Lab. Clin. Med.* **1997**, *130*, 436–441. [CrossRef]
2. Windmiller, J.R.; Wang, J. Wearable electrochemical sensors and biosensors: A review. *Electroanalysis* **2013**, *25*, 29–46. [CrossRef]
3. Tortorich, R.P.; Shamkhalichenar, H.; Choi, J.W. Inkjet-printed and paper-based electrochemical sensors. *Appl. Sci.* **2018**, *8*, 288. [CrossRef]
4. Moreira, F.T.; Ferreira, M.J.M.; Puga, J.R.; Sales, M.G.F. Screen-printed electrode produced by printed-circuit board technology. Application to cancer biomarker detection by means of plastic antibody as sensing material. *Sens. Actuators B Chem.* **2016**, *223*, 927–935. [CrossRef] [PubMed]

5. Radio Proximity (VT) Fuzes. Available online: https://www.history.navy.mil/research/library/online-reading-room/title-list-alphabetically/r/radio-proximty-vt-fuzes.html (accessed on 6 November 2017).
6. LaDou, J. Printed circuit board industry. *Int. J. Hyg. Environ. Health* **2006**, *209*, 211–219. [CrossRef]
7. Chan, C.M.; Tong, K.H.; Leung, S.L.; Wong, P.S.; Yee, K.W.; Bayes, M.W. Development of novel immersion gold for electroless nickel immersion gold process (ENIG) in PCB applications. In Proceedings of the 2010 5th International Microsystems Packaging Assembly and Circuits Technology Conference, Taipei, Taiwan, 20–22 October 2010; pp. 1–4.
8. Ratzker, M.; Pearl, A.; Osterman, M.; Pecht, M.; Milad, G. Review of capabilities of the ENEPIG surface finish. *J. Electron. Mater.* **2014**, *43*, 3885–3897. [CrossRef]
9. Pei, X.; Kang, W.; Yue, W.; Bange, A.; Heineman, W.R.; Papautsky, I. Disposable copper-based electrochemical sensor for anodic stripping voltammetry. *Anal. Chem.* **2014**, *86*, 4893–4900. [CrossRef]
10. Goyal, D.; Lane, T.; Kinzie, P.; Panichas, C.; Kam Meng, C.; Villalobos, O. Failure mechanism of brittle solder joint fracture in the presence of electroless nickel immersion gold (ENIG) interface. In Proceedings of the 52nd Electronic Components and Technology Conference, San Diego, CA, USA, 28–31 May 2002; pp. 732–739.
11. Accogli, A.; Lucotti, A.; Magagnin, L. In Situ-Raman spectroscopy and electrochemical characterization on electroless nickel immersion gold process. *ECS Trans.* **2017**, *75*, 1–6. [CrossRef]
12. Anastasova, S.; Kassanos, P.; Yang, G.Z. Multi-parametric rigid and flexible, low-cost, disposable sensing platforms for biomedical applications. *Biosens. Bioelectron.* **2018**, *102*, 668–675. [CrossRef]
13. Dutta, G.; Regoutz, A.; Moschou, D. Commercially fabricated printed circuit board sensing electrodes for biomarker electrochemical detection: The importance of electrode surface characteristics in sensor performance. *Multidiscip. Digit. Publ. Inst. Proc.* **2018**, *2*, 741. [CrossRef]
14. Evans, D.; Papadimitriou, K.I.; Vasilakis, N.; Pantelidis, P.; Kelleher, P.; Morgan, H.; Prodromakis, T. A novel microfluidic point-of-care biosensor system on printed circuit board for cytokine detection. *Sensors* **2018**, *18*, 4011. [CrossRef]
15. Kim, K.Y.; Chang, H.; Lee, W.D.; Cai, Y.F.; Chen, Y.J. The influence of blood glucose meter resistance variation on the performance of a biosensor with a gold-coated circuit board. *J. Sens.* **2019**, *2019*, 5948182. [CrossRef]
16. Chow, K.M.; Ng, W.Y.; Yeung, L.K. Barrier properties of Ni, Pd and Pd-Fe for Cu diffusion. *Surf. Coat. Technol.* **1998**, *105*, 56–64. [CrossRef]
17. Bozkurt, A.; Lal, A. Low-cost flexible printed circuit technology–based microelectrode array for extracellular stimulation of the invertebrate locomotory system. *Sens. Actuators A Phys.* **2011**, *169*, 89–97. [CrossRef]
18. Li, X.; Zang, J.; Liu, Y.; Lu, Z.; Li, Q.; Li, C.M. Simultaneous detection of lactate and glucose by integrated printed circuit board based array sensing chip. *Anal. Chim. Acta* **2013**, *771*, 102–107. [CrossRef]
19. Faria, A.M.; Peixoto, E.B.M.I.; Adamo, C.B.; Flacker, A.; Longo, E.; Mazon, T. Controlling parameters and characteristics of electrochemical biosensors for enhanced detection of 8-hydroxy-2'-deoxyguanosine. *Sci. Rep.* **2019**, *9*, 7411. [CrossRef] [PubMed]
20. Kassanos, P.; Anastasova, S.; Yang, G. A low-cost amperometric glucose sensor based on PCB technology. In Proceedings of the 2018 IEEE SENSORS Conference, New Delhi, India, 28–31 October 2018.
21. Tseng, H.Y.; Adamik, V.; Parsons, J.; Lan, S.S.; Malfesi, S.; Lum, J.; Shannon, L.; Gray, B. Development of an electrochemical biosensor array for quantitative polymerase chain reaction utilizing three-metal printed circuit board technology. *Sens. Actuators B Chem.* **2014**, *204*, 459–466. [CrossRef]
22. Nandakumar, V.; Bishop, D.; Alonas, E.; LaBelle, J.; Joshi, L.; Alford, T.L. A low-cost electrochemical biosensor for rapid bacterial detection. *IEEE Sens. J.* **2011**, *11*, 210–216. [CrossRef]
23. Bhavsar, K.; Fairchild, A.; Alonas, E.; Bishop, D.K.; La Belle, J.T.; Sweeney, J.; Alford, T.L.; Joshi, L. A cytokine immunosensor for multiple sclerosis detection based upon label-free electrochemical impedance spectroscopy using electroplated printed circuit board electrodes. *Biosens. Bioelectron.* **2009**, *25*, 506–509. [CrossRef]
24. György, I. Pseudo-reference electrodes. In *Handbook of Reference Electrodes*; Inzelt, G., Lewenstam, A., Scholz, F., Eds.; Springer: Berlin/Heidelberg, Germany, 2013; pp. 331–332. [CrossRef]
25. Moschou, D.; Trantidou, T.; Regoutz, A.; Carta, D.; Morgan, H.; Prodromakis, T. Surface and electrical characterization of Ag/AgCl pseudo-reference electrodes manufactured with commercially available PCB technologies. *Sensors* **2015**, *15*, 18102–18113. [CrossRef]
26. Raza, W.; Ahmad, K. A highly selective Fe@ ZnO modified disposable screen printed electrode based non-enzymatic glucose sensor (SPE/Fe@ ZnO). *Mater. Lett.* **2018**, *212*, 231–234. [CrossRef]

27. Rungsawang, T.; Punrat, E.; Adkins, J.; Henry, C.; Chailapakul, O. Development of electrochemical paper-based glucose sensor using cellulose-4-aminophenylboronic acid-modified screen-printed carbon electrode. *Electroanalysis* **2016**, *28*, 462–468. [CrossRef]
28. Abellán-Llobregat, A.; Jeerapan, I.; Bandodkar, A.; Vidal, L.; Canals, A.; Wang, J.; Morallon, E. A stretchable and screen-printed electrochemical sensor for glucose determination in human perspiration. *Biosens. Bioelectron.* **2017**, *91*, 885–891. [CrossRef]
29. Bihar, E.; Wustoni, S.; Pappa, A.M.; Salama, K.N.; Baran, D.; Inal, S. A fully inkjet-printed disposable glucose sensor on paper. *Npj Flex. Electron.* **2018**, *2*, 1–8. [CrossRef]
30. Romeo, A.; Moya, A.; Leung, T.S.; Gabriel, G.; Villa, R.; Sánchez, S. Inkjet printed flexible non-enzymatic glucose sensor for tear fluid analysis. *Appl. Mater. Today* **2018**, *10*, 133–141. [CrossRef]
31. Bernasconi, R.; Mangogna, A.; Magagnin, L. Low cost inkjet fabrication of glucose electrochemical sensors based on copper oxide. *J. Electrochem. Soc.* **2018**, *165*, B3176–B3183. [CrossRef]
32. Xuan, X.; Yoon, H.S.; Park, J.Y. A wearable electrochemical glucose sensor based on simple and low-cost fabrication supported micro-patterned reduced graphene oxide nanocomposite electrode on flexible substrate. *Biosens. Bioelectron.* **2018**, *109*, 75–82. [CrossRef]
33. Ribet, F.; Stemme, G.; Roxhed, N. Ultra-miniaturization of a planar amperometric sensor targeting continuous intradermal glucose monitoring. *Biosens. Bioelectron.* **2017**, *90*, 577–583. [CrossRef]
34. Buk, V.; Pemble, M.E. A highly sensitive glucose biosensor based on a micro disk array electrode design modified with carbon quantum dots and gold nanoparticles. *Electrochim. Acta* **2019**, *298*, 97–105. [CrossRef]
35. Cash, K.J.; Clark, H.A. Nanosensors and nanomaterials for monitoring glucose in diabetes. *Trends Mol. Med.* **2010**, *16*, 584–593. [CrossRef]
36. Power, A.C.; Gorey, B.; Chandra, S.; Chapman, J. Carbon nanomaterials and their application to electrochemical sensors: A review. *Nanotechnol. Rev.* **2018**, *7*, 19–41. [CrossRef]
37. Wang, J. Carbon-nanotube based electrochemical biosensors: A review. *Electroanalysis* **2005**, *17*, 7–14. [CrossRef]
38. Alhans, R.; Singh, A.; Singhal, C.; Narang, J.; Wadhwa, S.; Mathur, A. Comparative analysis of single-walled and multi-walled carbon nanotubes for electrochemical sensing of glucose on gold printed circuit boards. *Mater. Sci. Eng. C* **2018**, *90*, 273–279. [CrossRef]
39. Luo, X.L.; Xu, J.J.; Du, Y.; Chen, H.Y. A glucose biosensor based on chitosan–glucose oxidase–gold nanoparticles biocomposite formed by one-step electrodeposition. *Anal. Biochem.* **2004**, *334*, 284–289. [CrossRef] [PubMed]
40. Pu, Z.; Wang, R.; Wu, J.; Yu, H.; Xu, K.; Li, D. A flexible electrochemical glucose sensor with composite nanostructured surface of the working electrode. *Sens. Actuators B Chem.* **2016**, *230*, 801–809. [CrossRef]
41. Evans, D.; Papadimitriou, K.I.; Greathead, L.; Vasilakis, N.; Pantelidis, P.; Kelleher, P.; Morgan, H.; Prodromakis, T. An assay system for point-of-care diagnosis of tuberculosis using commercially manufactured PCB technology. *Sci. Rep.* **2017**, *7*, 685. [CrossRef]
42. Dutta, G.; Jallow, A.A.; Paul, D.; Moschou, D. Label-free electrochemical detection of S. mutans exploiting commercially fabricated printed circuit board sensing electrodes. *Micromachines* **2019**, *10*, 575. [CrossRef] [PubMed]
43. Wang, L.; Xue, L.; Guo, R.; Zheng, L.; Wang, S.; Yao, L.; Huo, X.; Liu, N.; Liao, M.; Li, Y.; et al. Combining impedance biosensor with immunomagnetic separation for rapid screening of Salmonella in poultry supply chains. *Poult. Sci.* **2020**, *99*, 1606–1614. [CrossRef] [PubMed]
44. La Belle, J.T.; Shah, M.; Reed, J.; Nandakumar, V.; Alford, T.L.; Wilson, J.W.; Nickerson, C.A.; Joshi, L. Label-free and ultra-low level detection of Salmonella enterica Serovar Typhimurium using electrochemical impedance spectroscopy. *Electroanalysis* **2009**, *21*, 2267–2271. [CrossRef]
45. Moschou, D.; Greathead, L.; Pantelidis, P.; Kelleher, P.; Morgan, H.; Prodromakis, T. Amperometric IFN-γ immunosensors with commercially fabricated PCB sensing electrodes. *Biosens. Bioelectron.* **2016**, *86*, 805–810. [CrossRef]
46. Umek, R.M.; Lin, S.W.; Vielmetter, J.; Terbrueggen, R.H.; Irvine, B.; Yu, C.J.; Kayyem, J.F.; Yowanto, H.; Blackburn, G.F.; Farkas, D.H.; et al. Electronic detection of nucleic acids: A versatile platform for molecular diagnostics. *J. Mol. Diagn.* **2001**, *3*, 74–84. [CrossRef]

47. Gassmann, S.; Götze, H.; Hinze, M.; Mix, M.; Flechsig, G.; Pagel, L. PCB based DNA detection chip. In Proceedings of the IECON 2012—38th Annual Conference on IEEE Industrial Electronics Society, Montreal, QC, Canada, 25–28 October 2012.
48. Sánchez, J.L.A.; Henry, O.Y.F.; Joda, H.; Solnestam, B.W.; Kvastad, L.; Johansson, E.; Akan, P.; Lundeberg, J.; Lladach, N.; Ramakrishnan, D.; et al. Multiplex PCB-based electrochemical detection of cancer biomarkers using MLPA-barcode approach. *Biosens. Bioelectron.* **2016**, *82*, 224–232. [CrossRef] [PubMed]
49. Jacobs, M.; Muthukumar, S.; Panneer Selvam, A.; Engel Craven, J.; Prasad, S. Ultra-sensitive electrical immunoassay biosensors using nanotextured zinc oxide thin films on printed circuit board platforms. *Biosens. Bioelectron.* **2014**, *55*, 7–13. [CrossRef]
50. Kuss, S.; Amin, H.M.A.; Compton, R.G. Electrochemical detection of pathogenic bacteria—recent strategies, advances and challenges. *Chem. Asian J.* **2018**, *13*, 2758–2769. [CrossRef]
51. Srivastava, S.K.; Van Rijn, C.J.; Jongsma, M.A. Biosensor-based detection of tuberculosis. *RSC Adv.* **2016**, *6*, 17759–17771. [CrossRef]
52. Kirk, M.D.; Pires, S.M.; Black, R.E.; Caipo, M.; Crump, J.A.; Devleesschauwer, B.; Döpfer, D.; Fazil, A.; Fischer-Walker, C.L.; Hald, T. World health organization estimates of the global and regional disease burden of 22 foodborne bacterial, protozoal, and viral diseases 2010: A data synthesis. *PLoS Med.* **2015**, *12*, e1001921. [CrossRef]
53. Farkas, D.H. Bioelectric detection of DNA and the automation of molecular diagnostics. *JALA J. Assoc. Lab. Autom.* **1999**, *4*, 20–24. [CrossRef]
54. Liu, R.H.; Yang, J.; Lenigk, R.; Bonanno, J.; Grodzinski, P. Self-contained, fully integrated biochip for sample preparation, polymerase chain reaction amplification, and DNA microarray detection. *Anal. Chem.* **2004**, *76*, 1824–1831. [CrossRef]
55. Lian, K.; O'Rourke, S.; Sadler, D.; Eliacin, M.; Gamboa, C.; Terbrueggen, R.; Chason, M. Integrated microfluidic components on a printed wiring board platform. *Sens. Actuators B Chem.* **2009**, *138*, 21–27. [CrossRef]
56. Lewicka, Z.; Yu, W.; Colvin, V. An alternative approach to fabricate metal nanoring structures based on nanosphere lithography. In Proceedings of the SPIE 8102, Nanoengineering: Fabrication, Properties, Optics, and Devices VIII, San Diego, CA, USA, 23 September 2011.
57. Vasilakis, N.; Papadimitriou, K.I.; Evans, D.; Morgan, H.; Prodromakis, T. The Lab-on-PCB framework for affordable, electronic-based point-of-care diagnostics: From design to manufacturing. In Proceedings of the 2016 IEEE Healthcare Innovation Point-Of-Care Technologies Conference (HI-POCT), Cancun, Mexico, 9–11 November 2016.
58. Moschou, D.; Tserepi, A. The lab-on-PCB approach: Tackling the μTAS commercial upscaling bottleneck. *Lab Chip* **2017**, *17*, 1388–1405. [CrossRef]
59. Hafler, D.A.; Weiner, H.L. Immunologic mechanisms and therapy in multiple sclerosis. *Immunol. Rev.* **1995**, *144*, 75–107. [CrossRef]
60. Lee, T.; Lee, Y.; Park, S.Y.; Hong, K.; Kim, Y.; Park, C.; Chung, Y.H.; Lee, M.-H.; Min, J. Fabrication of electrochemical biosensor composed of multi-functional DNA structure/Au nanospike on micro-gap/PCB system for detecting troponin I in human serum. *Colloids Surf. B Biointerfaces* **2019**, *175*, 343–350. [CrossRef]
61. Temiz, Y.; Lovchik, R.D.; Kaigala, G.V.; Delamarche, E. Lab-on-a-chip devices: How to close and plug the lab? *Microelectron. Eng.* **2015**, *132*, 156–175. [CrossRef]
62. Yamada, K.; Choi, W.; Lee, I.; Cho, B.-K.; Jun, S. Rapid detection of multiple foodborne pathogens using a nanoparticle-functionalized multi-junction biosensor. *Biosens. Bioelectron.* **2016**, *77*, 137–143. [CrossRef] [PubMed]
63. Nikkhoo, N.; Cumby, N.; Gulak, P.G.; Maxwell, K.L. Rapid bacterial detection via an all-electronic CMOS biosensor. *PLoS ONE* **2016**, *11*, e0162438. [CrossRef]
64. Yun, K.S.; Gil, J.; Kim, J.; Kim, H.J.; Kim, K.; Park, D.; Kim, M.s.; Shin, H.; Lee, K.; Kwak, J.; et al. A miniaturized low-power wireless remote environmental monitoring system based on electrochemical analysis. *Sens. Actuators B Chem.* **2004**, *102*, 27–34. [CrossRef]
65. Zhao, C.; Thuo, M.M.; Liu, X. A microfluidic paper-based electrochemical biosensor array for multiplexed detection of metabolic biomarkers. *Sci. Technol. Adv. Mater.* **2013**, *14*, 054402. [CrossRef]
66. Beni, V.; Nilsson, D.; Arven, P.; Norberg, P.; Gustafsson, G.; Turner, A.P.F. Printed electrochemical instruments for biosensors. *ECS J. Solid State Sci. Technol.* **2015**, *4*, S3001–S3005. [CrossRef]

67. Kim, J.; Sempionatto, J.R.; Imani, S.; Hartel, M.C.; Barfidokht, A.; Tang, G.; Campbell, A.S.; Mercier, P.P.; Wang, J. Simultaneous monitoring of sweat and interstitial fluid using a single wearable biosensor platform. *Adv. Sci.* **2018**, *5*, 1800880. [CrossRef]
68. Shamkhalichenar, H.; Choi, J.W. An inkjet-printed non-enzymatic hydrogen peroxide sensor on paper. *J. Electrochem. Soc.* **2017**, *164*, B3101. [CrossRef]
69. Shamkhalichenar, H.; Choi, J.W. Non-enzymatic hydrogen peroxide electrochemical sensors based on reduced graphene oxide. *J. Electrochem. Soc.* **2020**, *167*, 037531. [CrossRef]
70. Chen, S.; Yuan, R.; Chai, Y.; Hu, F. Electrochemical sensing of hydrogen peroxide using metal nanoparticles: A review. *Microchim. Acta* **2013**, *180*, 15–32. [CrossRef]
71. George, J.M.; Antony, A.; Mathew, B. Metal oxide nanoparticles in electrochemical sensing and biosensing: A review. *Microchim. Acta* **2018**, *185*, 358. [CrossRef] [PubMed]
72. Beitollahi, H.; Mohammadi, S.Z.; Safaei, M.; Tajik, S. Applications of electrochemical sensors and biosensors based on modified screen-printed electrodes: A review. *Anal. Methods* **2020**, *12*, 1547–1560. [CrossRef]
73. Hwang, D.W.; Lee, S.; Seo, M.; Chung, T.D. Recent advances in electrochemical non-enzymatic glucose sensors–A review. *Anal. Chim. Acta* **2018**, *1033*, 1–34. [CrossRef]
74. Arya, S.K.; Estrela, P. Recent advances in enhancement strategies for electrochemical ELISA-based immunoassays for cancer biomarker detection. *Sensors* **2018**, *18*, 2010. [CrossRef]
75. Loo, S.W.; Pui, T.S. Cytokine and cancer biomarkers detection: The dawn of electrochemical paper-based biosensor. *Sensors* **2020**, *20*, 1854. [CrossRef]
76. Rafique, B.; Iqbal, M.; Mehmood, T.; Shaheen, M.A. Electrochemical DNA biosensors: A review. *Sensor Rev.* **2019**, *39*, 34–50. [CrossRef]

Publisher's Note: MDPI stays neutral with regard to jurisdictional claims in published maps and institutional affiliations.

© 2020 by the authors. Licensee MDPI, Basel, Switzerland. This article is an open access article distributed under the terms and conditions of the Creative Commons Attribution (CC BY) license (http://creativecommons.org/licenses/by/4.0/).

MDPI
St. Alban-Anlage 66
4052 Basel
Switzerland
Tel. +41 61 683 77 34
Fax +41 61 302 89 18
www.mdpi.com

Biosensors Editorial Office
E-mail: biosensors@mdpi.com
www.mdpi.com/journal/biosensors

www.ingramcontent.com/pod-product-compliance
Lightning Source LLC
LaVergne TN
LVHW070612100526
838202LV00012B/628